Timeless fermentations ℄

Author:

Dr. *Antonio Silvestro*, self-employed heavenly based.
Department of *'Mathematic, Physics and Natural Sciences'*, Ba *'Biological Sciences'* University Federico II of Naples, Naples (NA), 80055, Italy.
Department of *'Agriculture'*, MSc *'Plant, Food Science and Environmental Biotechnology'*, University Federico II of Naples, Portici (NA), 80100, Italy.

Abstract:

The present text would guide you in making your own fermented products such as dairy and beverages. Using natural, dripped from the mammary gland of the endless *Rhea's* domesticated animals and synthetic, generate by the resonating Goddess of the way from which her breast spilled, lactic herbal infusion, *Heracles* whey protein, yogurt, cheesecake and the best *Aristaeus* cheese emerging from *Apollo* and *Cyrene*. *Dionysus* and *Bacchus* asteroids influencing alcoholic beverages changing toward youth thanks to antioxidant polyphenols and anti-aging compounds like resveratrol in grapes (*Vitis vinifera*) acting on the telomerase enzymatic pathway in regulating the extension of the Deoxy Ribonucleic Acid (DNA).

Keywords: dairy, cheese, milk, whey, yogurt, enology, zymology, fermentation, wine, beer.

Correspondence for Copyright © permissions requirement to:
Dr. Antonio Silvestro born Friday 15th May 1992 at 20:00 under the Taurus sign ♉ ☐ ascendant Scorpio ♏ ☐ according the Greek specular to Monkey 猴 hóu rising dog (*Canis lupis*) Canis Major 狗 gǒu in Chinese and chestnut (*Castanea sativa*) in Druid, flower (*xochiti*) in Aztec astrology, North knot Capricorn ♑ ☐ goat (*Capra hircus*) 羊 yang and South knot Cancer ♋ ☐ rat (*Ractus norvegiensis*) 鼠 shǔ, resident in n°100 Nazario Sauro St., 80026, Casoria (NA) (Italia), number phone: +39 3382634244, emails: dr.antoniosilvestro@gmail.com, tonysilverxxx@gmail.com and antonio.silvestro5@studenti.unina.it.

'Metagenomics – A pandemic theoretical and practical review on environmental genomics for basic research and applied industrial biotechnology' © Copyright Antonio Silvestro, 2019

Index

- **Metagenomics**
- **Saturn in Capricorn (lactic) fermentations:**
1. Dairy livestock………………………………………………………………..…………2
2. Rhea milk……………………………………………………………………..………..8
3. Heracles serum………………………………………………………………………...10
4. Alcmene synthetic milk…………………………………………………………………11
5. Alveolar milk……………………………………………………………………………15
6. Aristaeus cheese………………………………………………………………………..19
7. Cheesecake……………………………………………………………………………..23
8. Yogurt…………………………………………………………………………………..23
9. Ice cream………………………………………………………………………….…...25
- Ethanolic fermentation - Dionysus/Bacchus the spirit of enology and zymology………26
- Acetic fermentation – Vinegar………………………………………………………….30
- Mercury in Virgo distillation……………………………………………………………47

Metagenomics

Microbes are ubiquitarian through the biosphere making a change continuously! Antoni van Leeuwenhoek, invented the first microscope to resolve them, permitting Louis Pasteur to classify some of them (*Pasteurellacae spp.*) protocolling a method for inhibiting their growth preventing food spoilage later utilized also in surgery chamber, inoculating with his homonymous pipette the first vaccine for chicken (*Gallus gallus*) cholera. **Metagenomics**, all set of a gene in a *medium*, in many different matrixes from which is possible isolate the microbiota and transfer it *in vitro* like your own mind, in which the Red Blood Cells (RBCs) enhance the induction of the human toroidal magnetic field, otherwise, to predict their growth and development through a culture-independent approach where it is feasible to reproduce the environment condition. Actually, has been estimated that the 99.9 % of the microorganisms is not readily culturable (1).

The identification of microbial species in a community of many populations can passing through *in situ* DNA extraction, 16S rRNA amplicon of variable regions, DGGE, band blotting and bioinformatic alignment to close reference (e.g., BLAST) or *de novo*. Genetic diversity of microbial consortia can be asses with Polymerase Chain Reaction (PCR) amplifying fragments of 16S rRNA genes by Denaturing Gradient Gel Electrophoresis (DGGE) breeding, for example, 1% of a population of sulphate-reducing bacteria with an oligonucleotide variable region V3 16S rRNA probe (2). Random Shotgun sequencing of DNA has been used for natural acidophilic biofilm of uncultivable microorganisms such as *Leptospirillum* and *Ferroplasma II*, founding that Single Nucleotide Polyphormisims (SNPs) are the main heterogeneous genomic biomarkers for strain typing (3). Intergenic Spacer Region (ISR) polymorphism has been used for discerning between many *Staphylococcus* species like *S. sciuri, S. haemolyticus, S. hominis, S. auricularis, S. condimenti, S. kloosi, S. vitulus, S. succinus, S. pasteuri, S. capitis* in Italian fermented sausages (5). Transconjugants strains of *Lactococcus lactis* have been utilized for gen transfer assay with direct plate conjugation, filter mating and mating on milk agar, finding this last technique as the best (6). A Mexican *Zea mays* beverage, Pozol, have been investigated with a polyphasic approach combining culture media, DGGE fingerprinting and fermentation metabolites showing a microbial succession from epiphytic aerobic passing through heterolactic (LAB), *Lactobacillus*

'Metagenomics – A pandemic theoretical and practical review on environmental genomics for basic research and applied industrial biotechnology' © Copyright Antonio Silvestro, 2019

fermentum, till homofermentative Lactic Acid Bacteria (LAB), *L. plantaru, L. delbruekii, L. casei* acidify continuously the dough (7). Diversity and population dynamics of baker products, such as *Secale cereale* flour and bran have been asses using PCR-DGGE finding clusters of different *Lactobacillus* species (8). *Secale cereale, Triticum aestivum, T. durum* (Rye and wheat) flour sourdough have been asses by 16 S rRNA gene pyrosequencing. Viable counts, o LABs, LAB:yeasts, rate of acidification, OTUs, one week sourdough propagation utilized finding in the flour contamination of *Proteobacteria (Acinetobacter,Pantoea, Pseudomonas, Comamonas, Enterobacter, Erwinia and Sphingomonas)* detecting active metabolites or *Bacterioidetes (Chryseobacterium)*, inhibited after 1st day propagation, except for *Eneterobacteriaceae*, becoming the phylum *Firmicutes* dominant from the 2nd day, *Weisella spp.* and the last day succession (10th) *Lactobacillus sakei. L. lactis* from the 1st till the 5th day in the tetraploid durum flour. The only yeast in all the growth curve was *Saccharomyces cerevisiae* (9). PCR-DGGE analysis, using V6-V8 of the 16S rRNA gene, have been conducted for monitoring the structure and for the identification of starter microbial communities during Ragusano artisanal Sicilian cheese manufacture finding in raw milk *Macrococcus caseolitycus, Lactococcus lactis, Leuconostoc meseneroides* while in the ripe cheese *Streptococcus thermophilus, Lactobacillus fermentum, L. delbruekii, L. casei.* (10). DNA extraction Grana Padano cheese *in situ* showed late blowing symptoms due to *Clostridium spp.* also isolate from Reinforced Clostridial Medium plate were concentration of butyric acid more than 100 mg/kg have been detected in the DGGE positives (11). In Stilton cheese have been detected with Fluorescence In Situ Hybridization (FISH). *Lactococcus lactis, Lactobacillus plantarum, Leuconostoc mesenteroides*, revealing that the dairy matrix is subdivided in a *Lactococccus spp.* core, LAB "underind" and *Leuconostoc spp.* ubiquity (12). LAB have been found amplifying 16S-23S rDNA inter-spacer and/or 16S rDNA in Mozzarella cheese (13); mesophilic and thermophilic LAB grown on Man Rogosa Sharp (MRS) and Rogosa agar at 30°C showed high viable counts and the highest diversity (14). VOC measured with GC/MS raised from the curd Caserta and Salerno (Campania region) have been shown to different flavour profile represented with Principal Component Analysis (PCA) and Multidimensional Scaling (MDS) capable to track the hand-cutted soft cheese authenticity. Biotypes of *Streptococcus thermophilus* have been revealed with lacS-PCR-DGGE in these cities. Purification of Caciocavallo Silano PCR amplicons have been done with High-Pressure Liquid Chromatography (HPLC) evaluated in TriEthylAmmonioum Acetate (TEAA) showing a Natural Whey Culture (NWC) Chromatogram unusable for discriminate the microbiota in different regions of South Italy due to a wide distribution of low similarities. Salerno (S) and Caserta (C) mozzarella microbiota have been compared among the phases finding definitive differences such as *Lactococcus spp.* high abundance in C raw milk (Lc) while *Streptococcus macedonicus* in the Ls, the presence of *Lactobacillus helveticus* in the NWC and Curd (Cs) in Salerno, *Lactobacillus kefiranofacien* in the Salerno Mozzarella (Ms) (15, 16). Pyrosequencing with Roche 454 community growing on a solid substrate, viable counting in Colony Forming Units (CFU) or liquid in Most Probable Number (MPN), have been extracted the 16S rDNA amplified, revealed with PCR-DGGE, blotted, sequenced and aligned with BLAST. Culture-dependent approach permitting to identify the dominant colony with the highest dilution, emerging variants in generally cultivable strains (e.g., 10^8 CFU *Streptococcus thermophilus* and 10^4 *Lactococcus lactic* in Mozzarella cheese). Polyphasic PCR-DGGE technique revealed by the analysis of DNA directly extracted *in situ* from selected NWC *Streptococcus thermophilus, Lactococcus lactis, Lactobacillus delbrueckii and Lactobacillus crispatus*, while two *Lactobacillus fermentum* and *Enterococcus faecalis* were identified by analyzing DNA from cultivable communities (17). Quality labels like the Protected Designation of Origin (PDO) describing the geographical environment of provenance are governed by the Regolamentation 2081/92 EEC. 144 cheese type in the 28-EU countries, reaching $40*10^3$ tonnes in France. 183 samples of the Swiss Emmental cheeses were combined by multivariate statistical analysis obtaining discriminant analysis (DA) and an Artificial Neural Network (ANN), similar to 95% and 91%, respectively. Infrared and front-face fluorescent spectroscopy have been permitted direct chemical fingerprinting (18). PCR-DGGE brought to surface labelling frauds of lyophilized probiotics yogurts preparations undeclaring *Bacillus* and *Bifidobacterium* species, specifically depending on V2–V3 16S rDNA variability (19,20). The last eubacteria genre in a community within *B. coryneforme* and *B.*

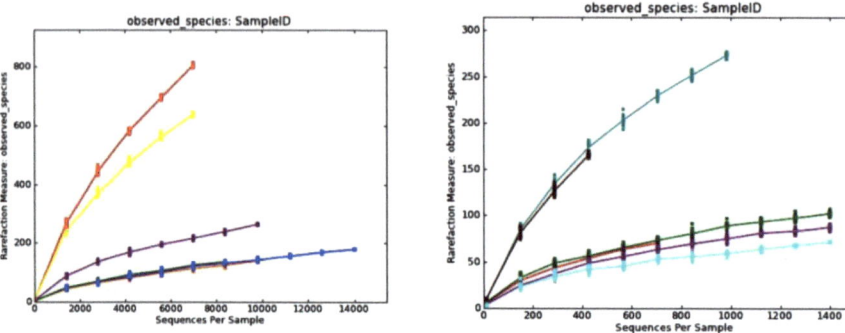

inducum population have been monitored during 24h-span development with nested-PCR-DGGE (21).

Figure 1 a, b Alpha diversity graphs Studying the microbiota through amplification of hyper variables regions (V1-V9) of stable 16S rRNA with primer-pair to calculate the diversity within an ecosystem (alpha diversity), via many indices Chao, Shannon, Good/ESC, assessing that it decreases through the process, actually the Operational Taxonomic Units (OTUs) increasing second a square root function with the increasing of reads number [e.g., from raw milk (6000 reads;800 OTUs) to the ripened cheese (14000 reads; 200 OTUs) "alpha diversity decrease almost 4-fold", while from fresh meat (1000 reads; 275 OTUs) to spoiled meat (1400 reads; 100 OTUs) " alpha diversity decreasing through the trend in linear with the increasing of the reads number].

Diversity inter-ecosystems (Beta diversity), using the bioinformatic tool Unifrac, multivariate statistical analyses (Hierarchical Clustering HC or Principal Coordinates Analyses PCoA) that can be represented with Unicfracs plots: weighted which does take into account differences in abundance of taxa between samples or unweighted which only considers the presence/absence of taxa between sample pairs. Ultra-high-throughput microbial community can be performed on Illumina HiSeq2000 platform generating up to 600 Gb of paired-end 10 base reads in a ten-days run (22). Some example of culture-independent HTS targeting variable regions of 16S rRNA of food microbiota: American cool ship ale fermentation V4 circa 150 bp Illumina GAIIx Taxonomy Resolution (TR): genus, fermented soybean (doenjang) V1-V2 circa 300 bp 454 FLX TR: species, Kefir V1-V2 circa 300bp 454 GS20, Meju V1-V3 500bp 454 FLX TR: species, Mozzarella V1-V3 454 Junior 500 bp. Screening target gene taxonomic relevant to amplify with a Polymerase Chain Reaction (PCR) using primer Illumina adapter or index pair (I5 e I7). Electrophoresis to purify the amplicons, discriminating the primer-dimers that migrate deeper through the gel. Multiplexing: G=Reference genome= *Homo sapiens* Illumina HiSeq2500 Truseq SBS v3 kit 1.5 Gb Pair-end reads (2*125bp) C=57; Roche 454 1 Mbp reads (700bp) C=0.2. Mapping: name sample and metadata. **Categorical**: to clustered in two or more groups on the base of the same parameter (e.g., sex, height, pH, ripened/unripen, fresh/spoiled) - Principal Coordinate Analysis (PCoA); or **Continuous variables**: any value in the range (pH, aw, Temperature, age, days of ripening, metabolite dose, etc.) - Principal Component Analysis (PCA), Hierarchical Clustering (HC), Heat plot Spearman, Kendall, Pearson co-occurrence co-exclusion correlations. Demultiplexing: clustering sequences for each sample based on the index used in PCR. Fastq files: Header, sequences and quality score (Phred). Quality filtering: eliminate short (< 300 bp) and low-quality sequences. Denoising: pyrosequencing high error-rate in the homopolymer 454 Roche. OTU picking: cluster 97% similarity, elimination singletons chimera. Merge sequences that came from pair-end Illumina sequencing (e.g., ea-utilis, Flash). Centroid choose: the representative sequence in the cluster, the most abundant, the longest or radically reducing the needed computational power. Taxonomic assignation comparing centroid with reference sequences on the database. OTU table: plot to represent the microbiota quali-quantitative composition. Other than targeting rRNA is possible, maintaining the same process steps, choosing other target sequences among the ones with the highest heterogeneity between biotypes. Quantitatively monitoring *Streptococcus thermophilus* biotypes during curd fermentation through the sequence of lacS highlighted 28 gene variants in Mozzarella (M), Grana Padano (GP) and Parmigiano Reggiano (PR) revealing that GP and PR fermentation is driven by *L. delbrueckii* and *L. helveticus*, while *S.thermophilus* dominated in M (23). Minimum Entropy Decomposition (MED) algorithm for a sensitive oligotiping partitioning high similar sequences of HTS marker genes into MED nodes, homogeneous OTUs have been utilized to detect to gamma proteobacteria microbial community within *Hexadella compare with a confere dedritifera cryptic* sponge (24). Shotgun metagenomics analysis workflow: Pre-process size fragment samples with BioAnalyzer that has an output plot (size [bp]; Fluorescent Unit [FU]), filtering contaminant reads out (BMTagger, DeconSeq), trimming adaptors and short quality bases, filtering adaptor-dimers and reads out (CutAdapt, Fastx-toolkit, Adapter Removal), de-duplicating (Prinseq, SolexaQA++, Sequence Alignment Map tools). 1) Mapping: short-reads alignment (Bowtie, BWA), gene catalogues (MetaHit) Taxonomic classification binning the reads (MEGAN) or focus on phylogenetically informative genes (MetaPhlan, mOTU). *De-novo* assembly: functional profile, reconstruct genes from metagenomes assembly short-reads (Velvet, Oases, SOAP) making Scaffold with Contigs. Metabolomic data analysis with pathways databases (e.g., KEGG, Reactom, Panter, MetaCyc, CAZy, BRENDA). Pan-genomics: based assembly: BLAST, identify core 99-100 %, 15 < shell < 20 %. Short reads: StrainPhlan, PanPhlan. Evolution of the microbiological approaches used to study microbial diversity and ecological traits in food system begin from planting, passing through counting isolation, fingerprinting (e.g., DGGE) till rRNA-target metagenomics. Amplicon targeted high-throughput food meta-study as the "Food microbionet":

interactive web-based visualisation of Gephi graphs for an efficient exploration of microbial communities in dairy, meat, sourdough and fermented vegetables, detecting a general low complexity than human microbiome, soil and environmental matrices and significant non-random patterns of co-occurrence (copresence and mutual exclusion) in incidence and abundance data with CoNet bioinformatic tool (25). Depending on the pipeline used for taxonomy assignment Ribosomal Database Pipeline (Bacterial and Archaeal 16S rRNA and Fungal 28S rRNA sequences) or UCLUST can be reported low differences in the OTUs, but not significative, with a higher number of OTUs corresponding to a lower taxa richness. Bacterial biogeographical patterns in a cooking centre for hospital foodservice have been evaluated, detecting the potential contaminants such as *Pseudomonas, Psychrobacter, Paracoccus, or Kocuria*, through the assessment of the microbial consortium swab-sampling 16S rRNA culture-independent HTS. 500 OTUs extremely variable in their relative abundances (0.02-99%) depending on the specie, more than 70 % of the samples, ascribed to the *Acinetobacter, Chryseobacterium, Moraxellaceae*, and *Alicyclobacillus* counting below than 10% of all the community. The cleaning procedures, food, and storage sanitary method have been a high impact on the microbial community (26). Potential contamination has been found in dairy plant 16S rRNA and 26S rRNA based culture-independent HTS amplicon sequencing, 200 OTUs extremely variable in abundance (range between 0.01 and 99%. 70% of the sample that has been found was LAB, mainly *Streptococcus thermophilus*, and potential spoilage, *Pseudomonas, Acinetobacter, Psychrobacter*, while among the yeasts the most abundant were *Kluyveromyces marxianus, Yamadazima triangularis, Thrichosporon fecal* and *Debaryomyces hansenii*. The occurrence of spoilers excluded LAB (27). Despite the initial contamination of *Proteobacteria Pseudomonas, Firmicutes Streptococcaceae, Actinobacteria Propiobacteriaceae*, and *Bacteroidetes Flavobacteriaceae* have been found as the most abundant phylum related, the psychrotrophic *Firmicutes Leconostoc gelidum subsp. gasicomitatum* and *L. gelidum subsp. gelidum* were the most representative at the end of the anaerobic cold-storage, as dominant spoiler of ready-to-eat (RTE) Finnish, Belgian and Japanese (28a) meals, elucidated via viable count, HTS 16S rRNA and bioinformatics pipeline, mainly expressing genes involved in carbohydrates and amino acids metabolisms pathways (e.g. Glycolysis/Gluconeogenesis, pyruvate, lactose and pentose phosphate, fructose-mannose metabolisms). To note a strong co-occurrence between *Proteobacteria Xantomonadaceae* and *Methyloversatilis* and strong co-exclusion between *Firmicutes Staphylococcus* and *Bacillus* species (29). Meat spoilage, the pauperization of organoleptic and nutritional values of flesh, due to the physical and chemical (e.g., oxidation) apart from microorganisms. Among this last *Proteobacteria/Gammaproteobactiria/Pseudomonadales/Pseudomonas* spp. (some of the 24610 as reported on Nation Centre of Biotechnology Institute - Taxonomy), Firmicutes/*Bacilli/Bacillales/Listeriaceae/Brochotrix thermosfacta*, *Proteobacteria/Gammaproteobactiria/Enterobacteriales/Enterobacteriaceae* and *Firmicutes/Bacilli (LAB)*, can show different sigmoidal trends in Air (A), Vacuum (VP) and Modified Atmosphere packages (MAP) (T=0-5°C). In Air Pseudomonas spp. grow and developed themselves reaching the highest CFU/cm^2 at the asymptote – contamination threshold (10-12 h;7 CFU/cm^2) while in Vacuum envelope LAB, releasing acid odor, in just 6-12 hours, in 10-20h in MAP with no-off odor. As the *Liebing's Law of Minimum* describing just one resource as limiting factor in a multivariate matrix approximate far from realty, so one causing agent can be pursued exclusively as spoiler. Gram-positive bacteria LAB is associated with perishing meat contributing organoleptic downgrading, but can manifest also bioprotective action. In VP, VOCs can be utilized to odour-identify them, among the family *Lactobacillariaceae*: *L.sakei* spreads Aldehydes Heptanal (wine-lee, oily rancid), *L.algidus* Benzaldehyde (almond), *L.fuchuensis* 2-Butanone (camphor), *L.piscium* Volatile fatty acetic acid (vinegar), *L.curvatus* Methylthioacetate (eggy). While in the family *Leuconostocaceae* in which are ascribed *Leuconostoc mesenteroides, L.carnosum and L. gelidum subsp gasicomitatum*, this last for example diffuse, in A and/or VP, alcohols (Butanol-fruity), aldehydes (1-Hexanol-fusel oil, Hexanal-fresh green, Heptal, Nonanal-orange peels), Ketones

(Acetoin/3-hydroxyquinone, Diacetyl/butane-2,3-dione-dairy, gamy; 2-Butanone), VFA (acetate, butane-butter, hexanoate-butyrate). Eventually, in the *Carnobacteriaceae carnobacterium* divergence, but furthermore *C.maltoaromaticum* is characterized by a wide VOC profile (28b). In butcher retailers have been conduct according to the Regulation (EC) 2073/2005 diagnosis of carcass swabs, plants, end products - beef cuts (chunk, brisket, flank, loin, ribs, etc.) microbial community assessing the diversity of the most probable contaminants with a 16S rRNA amplicon culture-independent HTS, finding more than 600 OTUs in the first complex matrix decreasing through the processing (30). Small-scale (SD) and large-scale retailer distribution (LD) have been evaluated with relative distances - Unifrac and Phylogeny Investigation of Communities by Reconstruction of Unobserved States (PiCrust (31)), tool for predictive functional profiling using 16S rRNA, finding *Firmicutes* in environmental swabs related to carbohydrates, terpenoids and polyketides metabolisms, while *Proteobacteria* in the meat due to a rising of xenobiotics biodegradation, protein, and lipid catabolism, nevertheless the 80% of the core microbiota was shared (despite of for the carcasses – high α diversity and the differences) between the SD and LD wasn't significant (32). Sampling each week beef stored at 4°C through 45 days and then metabolite analysing spoilage microorganisms by solid-phase microextraction (SPME)-gas chromatography (GC)-mass spectrometry (MS) and proton nuclear magnetic resonance (1H NMR), and microbial diversity (PCR–DGGE) and pyrosequencing), leads for identifying soil (*Terrabacteria group*) rather than meat bacteria in VP and Antimicrobial Vacuum (AV), *B. thermosphacta* have been found in A (N_2 78 %, O_2 20 %, Ar 0.9%, CO_2 40 %) and MAP (O_2 60 %, CO_2 40%) during the first days of conservation, while LAB butanoic acid and *C. divergens* 1-octen-3-ol (mushroom, underwood) metabolizer showed resistance to bacteriocin in the AV (33). Combining MAP with nisin antimicrobial High-Density polyethylene (HDPE) film was an effective technique to control spoilage bacteria, retarding the growth of the main found *Carnobacterium spp.*, *B. thermosphacta*, *Pseudomonas fragi* and the *Gammaproteobacteria/Enterobacteriales/Yersiniaceae/Rahnella aquatilis* (34). In refrigerated meat the most abundant population detected was the *C. maltoaromaticum* characteraside by the peculiar presence of 2-ethy-1-hexanol, 2-buten-1-ol, 2 hexil-1-octanol, 2-nonanone, and 2-ethylhexanal and rich in aldehydes, lactones and sulphur compounds, while among just the mesophilic *Yersiniaceae/Serratia proteamaculans* were rich in alcohols (e.g. 1-octen-3-ol) and esters (e.g. isoamyl acetate) and only in the psychotrophic condition *P. fragi* with the highest number of alcohols and ketons within the about 100 VOCs founded with a Gas Chromatography- Mass Spectroscopy (GC/MS) (35). In this last gram-negative spoiler mentioned, grown in a *medium* containing citrate, haemoglobin, myoglobin iron chloride as iron source, in aerobic condition at 4 °C, sensory analysing with SPME/GC/MS have been found 4-methylthiophenol, 3-ocanone, ethyl hexanoate (pineapple, banana), ethyl octanoate (fruity wine, apricot, banana, brandy pear), ethylnonenoate, ethyl decanoate (apple, grape), 2-pentylfurane for more than 50% strains on the 65 typed on proteolytic and lipolitic activity (36). Metabolomics analysis of meat storage salted (NaCl) has been conducted with HeadSpace SPME finding *Carnobaterium maltoaromaticum* strains (e.g., D1203, L172) releasing dairy flavour compounds such as acetoin, 1-octen-3-ol and butanoic acid, more in A than V (37). As aforementioned, spoilage bio-typing, within a specie, is a complex biodynamic phenomenon of microbial associations that can be analysed combining omics techniques.

Saturn in Capricorn (lactic) fermentations:

1. Dairy livestock

Cow (*Bos taurus*) cattle livestock were *domesticated* in Mesopotamia during the Neolithic Revolution in the past 10.500 years both for beef and milk production. Currently, the main *producer* *countries* of milk are US, India and China. Average cow gestation time is nine months (t = 9 months) and the breed can reach m = 35 kg growing up to m = 2150 kg when adult, each accounting about 1000 €. These animals can perceive InfraSounds due to earthquakes and volcanoes. Cattle have one *stomach* with four compartments: the rumen, reticulum, the omasum and abomasum. Chewed food is known as *cud*, this undergoes metabolic digestive reactions via the microbes in the *rumen* both catabolic breaking down cellulose and other carbs into fuelling Short-Chain Fatty Acids (SCFA) or anabolic synthetizing amino acids from urea and ammonia. Daily, cows would wet feed m_{wet} = 50 kg and drink V_{water} = 120 L generating $V_{milk} \approx$ 25 L. The chromosome assortment of cows is 2n = 60.

About thirty breeds of *miniature cows* (h ≤ 110 cm, m ≤ 250 kg) are officially recognized, among them the beef and milk (4.5 L/day) producers Irish Dexter's, Herefords, Argentinian Angus, Australian Lowlines, Indian Kasagarod, US Zebus, Jersey, Dahomey, Indian Punganur and Vechur in needs of lees resources compared with the standard cows, precisely, m_{wet} = 5 kg/day and drink V_{water} = 12 L/day.

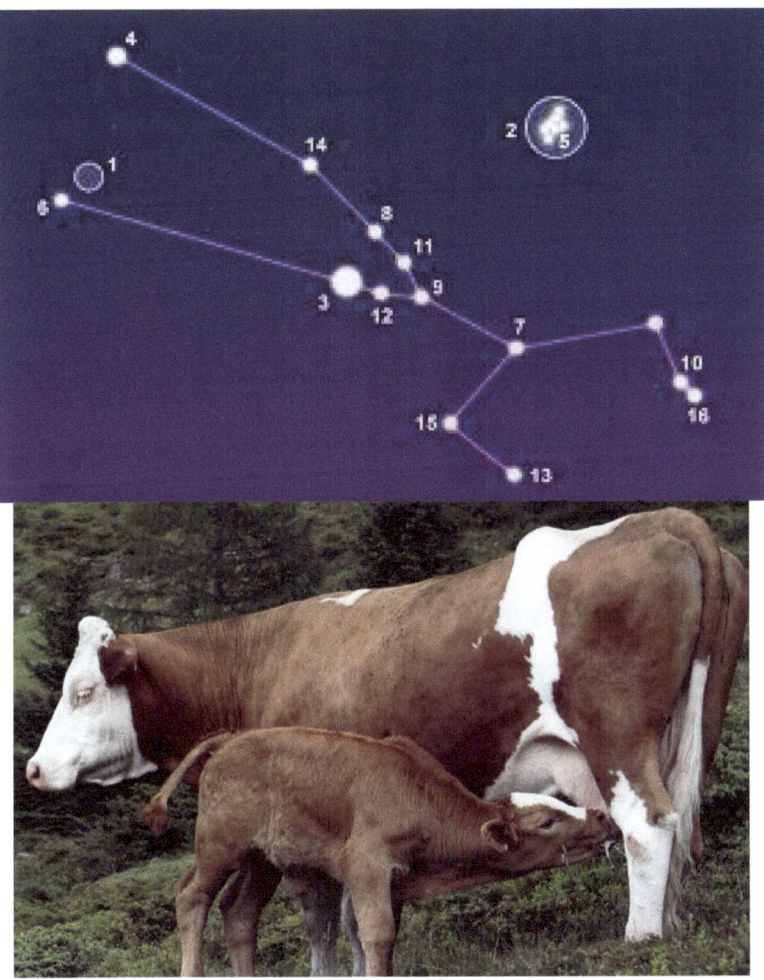

Figure 2 Taurus costellation and cow (*Bos taurus*) – Image source:
https://www.sciencenews.org/blog/growth-curve/does-breast-milk-come-pink-and-blue

Sheeps (*Ovis aries*) require slightly less space than goats, furthermore, this last tent to test their boundaries escaping the fences for this for beginner is preferential to choose the first. Regulating the soul of the ovine coherently with the radiation of the Saturn's moon Rhea (2nd largest after Titan, this even bigger than the planet Mercury), Greek Goddess of milk, managing their endocrine system administrating hormones is possible to induce lactation in non-pregnant mammals producing milk like the ewes, precisely, peptide somatropin (Growth Hormone GH) and prolactin pituitary hormones, secreted in response to the increased concentration of oxytocin and oestrogen naturally synthetized during the sexual estro period and menstrual progesterone inhibitors. Perhaps, let these animals listen binaural beats with earphones constantly would let them secrete milk from the mammary glands even if without lamb embryo in their placenta ($v_\delta \approx 1$ Hz, excluded $v_{progesterone}$ =1.446 Hz https://www.youtube.com/watch?v=AaN4PbJCpi4).

Figure 3 Aries constellation (top) and ovine **sheep** (*Ovis aries*, ram m_{max} = 160 kg, ewe m_{max} = 100 kg, lifetime = 11 years, reproductivity age = 7 months, milk production up to 6 years, bottom) used for their wool pelts, meat (50 % weight), milk that being double fat than cow and goat is preferentially skimmed from cream and is naturally higher in carbohydrate and proteins content compared to them suitable for cheese making (male ram 50 £, female ewe costs 45 £, but producing $180 < V_{milk} < 600$ L over a 3 months $< t_{lactation} < 8$ months/year, puppy wither 25 £ https://www.farmingads.co.uk/Sheep/A/livestock-for-sale.php?order=lowest, withers 50 € https://www.subito.it/animali/agnelli-biologalgfaici-napoli-356306813.htm) or (55 $ https://plancanada.ca/giftsofhope/product/35/goat) eat an average of 1 kg/day organic dry matter hay such as Ceres Graminaceae (wheat *Triticum spp.*, corn *Zea mays*, Fabaceae (alfalfa *Medicago sativa*, cowpea *Vigna unguiculata*, peanut *Arachis hypogaea*, soybean *Glycine max*, Timothy *Phleum pratense*, white and red clover/ladino *Trifolium spp.*), grasses (bermuda grass *Cynodon*, bromegrass *Bromus spp.*, Kentucky blue meadow grass *Poa pratensis*, reed canary grass *Phalaris arundinacea*, ryegrass *Lolium spp.*, tall fescue *Festuca arundinacea*, native and orchard grass), vaporized seed mix (22.5 €/30 kg https://www.millstore.it/mangime-fioccato-misto-4p-granfiocco#product-description), forb (not-graminoid herbaceous plant), fermented silage, pellet (https://www.gruppocarli.com/en/product-request-ok) and Amphitrite kelp algae (Kindle eBook 1.62 € https://www.amazon.com/dp/B08DL9B488).

Figure 4 Female doe and male buck domestic **goat** (*Capra aegagrus hircus*) (100 – 350 $ https://mountainflowerdairy.com/baby-goats-for-sale/, 80 $ https://plancanada.ca/giftsofhope/product/35/goat, 50 € https://www.subito.it/animali/caprette-tibetane-salerno-350850481.htm) – Image source: https://www.faidateingiardino.com/animali-da-compagnia/zootecnica/caprette-tibetane-allevarle-giardino

'Metagenomics – A pandemic theoretical and practical review on environmental genomics for basic research and applied industrial biotechnology' © Copyright Antonio Silvestro, 2019

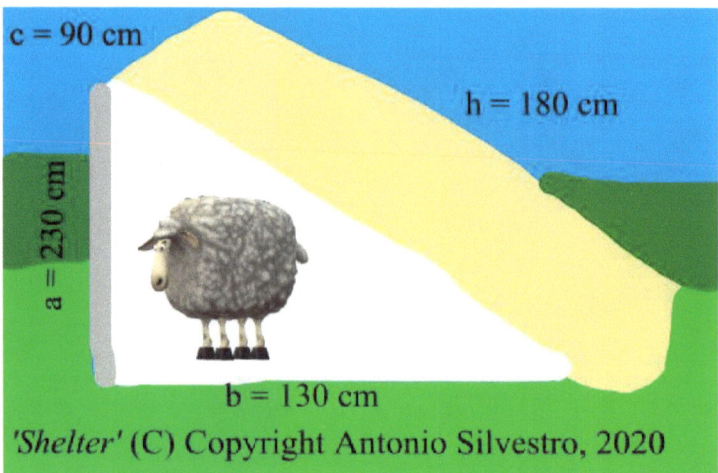

Figure 5 Shelter suitable for sheeps, goats and mini cows done of side fences (115 x 62 mm, 7 € https://www.wish.com/search/fences/product/5db2d59f88658a0b5282c716?source=search&position=92&share=web) and top diagonal sun shade canopy (180 x 130 x 230 mm, 12 €

https://www.wish.com/product/5b48043569017d1f0fc24a43?from_ad=goog_shopping&_display_country_code=IT&_force_currency_code=EUR&pid=googleadwords_int&c=%7BcampaignId%7D&ad_cid=5b48043569017d1f0fc24a43&ad_cc=IT&ad_lang=EN&ad_curr=EUR&ad_price=8.00&fallback_cids=5b1613f5a928cc2de1c62fd15b44206135c3be1e176d4373&campaign_id=8688966404&guest=true&gclid=Cj0KCQjwqfz6BRD8ARIsAIXQCf30W4CHafDgv6wTOEHhsaKaeb090Ta-cv19ociENjYAWGS2fxM7UWcaAqYPEALw_wcB&hide_login_modal=true&share=web) - Image source: (C) Copyright Antonio Silvestro, 2020.

Figure 6 **Donkeys** (*Equus africanus asinus*) reach at withers maximum length l = 160 cm, can live up to 50 years when well nourishing from mangers fille of healthy foraging feed *Fabaceae*: *Graminaceae* = 1 : 4, dry feed/1.5 % body weight/day. Minimum price 300 €.

2. Rhea milk

Figure 7 Rhea-Cybele/Ops statue of the goddess wife of Chronus/Saturn, related to its moon satellite or the 577 asteroids.

Cow **milk** (ρ_{milk} = 997 kg/m³ = 0.997 g/mL, pH < 6.6) is white, with some yellowish carotenoids reflexes due to the natural dyes within, bluish in the skimmed, greenish in the B2-enriched, colloidal emulsion of butterfat globules (0.2 < ø < 15 μm) water based (90 %) in which are dissolved organic macro- like 5 % carbohydrates (mostly disaccharide lactose $C_{12}H_{22}O_{11}$ = glucose + galactose), 4 % lipids of which 2/3 saturated, (TriAcylGlycerol TAG < 98%) and proteins (3 % 30g/L, ≈ 80 % αs1-, αs2-, β-, and κ-casein micelles $CaPO_4^{2-}$-bonded (ø$_{Casein}$ = 15 Å), mainly glutamic acid among the amino acids), ionized inorganics micro-nutrients 0.7 % 5 < [K, 120 mg Ca/100 g milk, P, Mg, Na, Cl] < 40 mM and vitamins (A, B1, B2, B3, B5, B6, B7, B9, B12, C, D, K, E), secreted by the mammary glands cells evolved by sweat apocrine *Permian-Triassic CynoDonta*, immune system leucocytes and symbionts microbiotas.

High Temperature Short Time (HTST), process of T = 72 °C for t = 15 s, has comparable sterilization efficiency with microfiltration. *Ultrafiltration*, instead works for isolation of lactose (ø$_{Lactose}$ = 3 mm) from the protenaiceous matrix suitable for cheese-making. Ultra Heat Treatment (UHT), at temperature T = 138 °C and time t = 2 s, prevent the growth and development of potential harmful bacteria such as *Listeria monocytogenes*. Milk UHT can be *stored* at room temperature up to t = 3 days, but in the fridge (T = 4 °C), despite labelled as expired, it can last up to t = 1 month and in the freezer (T = - 18 °C) up to t = 3 months, after which it will start to naturally acidify due to the native microbiome communities that were living in the mammal secreting it. Homogenization of the cluster of fat globules make the milk less prone to incomplete oxidation and/or fat hydrolyses, otherwise, rancidity, hence, increases the shelf life. Variable fat-content top-layer *cream* is made just standing milk for 12 < t < 24 h at room temperature, quickly make with centrifuges.

Cow emission ≈ 0.3 kg CO_2-C_{eq}/100 mL milk

H_2O : Cow Milk = 65 L : 100 mL - 1 L : Cow Milk = 0.9 m² : 100 mL - 1 Cow : milk 20 L/day

Milk is processed into a variety of *products* such as cream, butter, yogurt (breakfast m = 75 g), kefir, ice cream, whey protein, sparkling juices and cheese.

Infused milk: stir m = 250 g grinded dry herbs (also by-products) like *Melissa officinalis* with V = 1 L whole fatty milk for t = 1 h.

Infused pasteurized skim milk powder: double boiling V = 1 L infused milk for t = 2 h, dehydrator (or low heat oven) T = 50 °C and t = 10 h.

Temperature increment in t = 1 min for m = 1 kg of substance heated in *microwave* at power P = 750 W:

$$E = 750 \text{ W} \times 1 \text{ min} = 45 \text{ kJ}$$
$$750 \text{ J}/1 \text{ kg} = 45 \text{ J/g}$$
$$45 \text{ J/g} / 4.186 \text{ J/g °C} = 11 \text{ °C}$$
$$(1 \text{ °C} : 0.53 \text{ Wh} \rightarrow 50 \text{ °C} : 27 \text{ Wh})$$

3. Heracles serum

Heracles **whey** (= serum) is the opalescent liquid, pH ≤ 5, by-product of the cheese manufacturing resulting from the milk curdling [100 g : 100 kJ : C(H$_2$O) 5 g : Protein 0.85 g (*Bos taurus* 20 % whey 80 % casein)] containing lactose, lactalbumin (25 %), lactoglobulin (65 %), bovine serum albumin (8 %) and immunoglobulins, vitamins, minerals. Enzymatically Hydrolysate (WPH), high lactose carbohydrate Concentrates (WPC, 30 < proteins < 90 % dehydrate serum from soft cheese process), blend (WPIsolate Micro-Filter = 0.1 μm MF at T = 50°C), 1 tsp vanilla neither fat nor lactose *Isolate* (WPI, 90 % proteins). Body building shakes can be prepared dissolving isolated whey powder m = 50 g in water or 1 cheese serum : 3 H$_2$O.

Thermo-unstable proteins are properly isolate from their matrix and fragmented in their composing amino acids via differential pressure driven *dead-end filtration* through semi-permeable membranes of various porosity with the limiting one characterized by the diameter ø < 9 Å [øaa = -N-C = 2(0.37) + 2(0.7) + 2(0.77) + 2(0.72) + 2(0.67) ≈ 9 Å ('Atomic structures of all the twenty essential amino acids and a tripeptide, with bond lengths as sums of atomic covalent radii' by Raji Heyrovska, 2018)], that permit to the water molecules (ø$_{H20}$ = 2.75 Å = 2.7 : V$_{H20}$ = 10.89 Å^3) drip, leaving the amino acids, of averages molecular weight equal to 110 Da (= mg/mol), on top as the so called *'filter cake'*. Periodic washing cycles are needed for reducing the detrimental fouling increasing the life time of the bio-engineering system. The filtration dead-end ***Darcy's law*** would follow the equation:

$$Q = dV/dt = \Delta p/\mu \; A \; (1/R_f + R_s)$$

Where:
V = volume [L]
P = pressure [atm]
A = area [m^2]
Rf = retention of the fluid serum
Rs = retention of the solid cuddled
μ = viscosity [N s/m^2]

Permeable membrane technology for fluid separation or even particle accelerator can be used for breaking down the lipid's bonds.

Sparkling drink: 1 juice : 4 whey

It is to note that whey ca be substituted it to water for preparing high content **bakery** products and can be used for *Ceres* nitrogenated **fertilizers** dissolved it in 6 parts of water. Furthermore, a **sleeping herbal infusion** rich in casein can be made dissolving whey powder in warm milk and chamomile (*Matricaria officinalis*).

4. Alcmene synthetic milk

'Alcmene', the galactical synthetic milk from cow (*Bos taurus*), sheep (*Ovis aries*), goat (*Capra domestica*), donkeys (*Equus africanus asinus*) and even human (*Homo sapiens* and unofficial *Homo atm*) of the mother of *Heracles* rooted in the *Mulhadhara* for growing the *Sthula Sharir* of babies and not anymore.

Figure 8 Cute statue of the baby Heracles while holding Ophion with his hand, snake that ruled the Earth beside Eurynome, later confused in the salty water by the father Zeus that merry her, the reflection of his own insatiable intestine as in the 13rd costellation of the zodiac Ophiuchus.

An average *Bos taurus* would produce about V_{milk} = 21 L/day, but as the cattle would need a proper rural environment that would not fit with urban one, the *Electryone* synthetic milk would offer a worthy alternative for a home-made production with a relatively simple laboratory equipment suitable in all the kitchen.

TABLE 1
Assumed composition of milk

	GRAMS PER LITER	EQUIVALENTS PER LITER	MOLS PER LITER
K_2O	1.80	0.0382	
Na_2O	0.72	0.0232	
CaO	1.78	0.0635	
MgO	0.30	0.0148	
Total		0.1397	
P_2O_5	1.50		$0.0211(H_3PO_4)$
Citric	2.00		0.0104
Cl	1.00		0.0282
SO_2	0.11		0.0014
Total CO_2			0.0050
Casein	28.0		
Albumin	7.2		
Other protein	0.2		

Solution 1 { Mols
0.0074 MgO   0.298 grams
0.0211 KH_2PO_4 0.873 grams
0.0104 Citric acid. H_2O 2.185 grams } Dissolve with warming and dilute to 100 cc.

Solution 2 { 0.0014 $CaCO_3$ 0.140 grams
0.0014 H_2SO_4 28 cc. N/10 H_2SO_4 } Dissolve cold. Dilute to 50 cc. and use while fresh

Solution 3   0.0141 $CaCl_2$ 1.565 grams or preferably the equivalent of an analyzed solution. Dilute to 50 cc.

Solution 4 { 0.0171 KOH   17.1 cc. N/1 KOH
0.0232 NaOH 23.2 cc. N/1 NaOH } Dilute to 50 cc.

Solution 5 (for 200 cc. "synthetic milk" in which casein represents total protein)
7.0 grams casein
10.0 grams lactose } in 130 cc. N/20 $Ca(OH)_2$ solution

Chemical compound	Link selling company	Price [€]

'Metagenomics – A pandemic theoretical and practical review on environmental genomics for basic research and applied industrial biotechnology' © Copyright Antonio Silvestro, 2019

Compound	Source	Price
52 % K$_2$O flocculant + 34 % P$_2$O$_5$ = KH$_3$PO$_4$ (cardiomyocytes DNA)	https://www.aquasabi.com/warenkorb.php	17 €/500 g [8.73 g/L] => 57 L (0.3€/L)
Na$_2$O	https://www.chemicalbook.com/Price/SODIUM-OXIDE.htm	68 €/25 g [0.72 g/L] => 35 L (1.9 €/L)
CaO [fertilizer 2 (N 8 %, CaO 12 %)]	https://www.sigmaaldrich.com/catalog/product/aldrich/208159?lang=it&region=IT https://inscx.com/shop/product/calcium-oxide-nanopowder-cao-99-9-80nm/ https://www.manomano.it/catalogue/p/concime-nitrocal-l-1-kg-27306910?model_id=27287801).	48 € /25 g, 125units/£ 1.80, 9 €/kg [1.78 g/L] => 14 L (3.4€/L)
NaCl		$ 0.0042 /kg
SO$_3$ (piridine)	https://www.chemicalbook.com/Price/Sulfur-trioxide.htm https://www.fishersci.com/shop/products/sulfur-trioxide-pyridine-complex-acros-organics-4/AC132870250	145 $/40 g or 32 €/25 g [0.11 g/L] => 364 L (0.09 €/L)
CO$_2$ anhydrous (or gaseous from Dionysus ethanol fermentation)	https://www.doctorpoint.it/p1711-busta-ghiaccio-istantaneo-monouso-in-pe-18x15cm	0.38 + 14 €/18 x 15 cm
MgO (Chlorophyll)	https://www.iherb.com/pr/Now-Foods-Magnesium-Oxide-Pure-Powder-8-oz-227-g/695	10 €/240 g [2.98 g/L] => 80.5 L (0.124 €/L)
Citric acid (*Rutaceae* peels)	https://www.google.it/search?q=citric+acid+powder+buy&newwindow=1&sa=X&nfpr=1&biw=1366&bih=608&tbm=shop&sxsrf=ALeKk0317yE9guYDl4YG8rhLbnazr6LIYw:1594104285914&tbs=p_ord:p&ei=3RkEX-u1N_yEk74PkbKtsAU&ved=0ahUKEwirmMmdxbrqAhV8wsQBHRFZC1YQuw0I5wIoAQ#spd=6207314381384163378	7 €/227 g [21 g/L] => 11 L (0.636 €/L)
CaCO$_3$ (*Gallus gallus* egg shells)	https://www.iherb.com/pr/Now-Foods-Calcium-Carbonate-Powder-12-oz-340-g/480	11 €/340 g [2.8 g/L] => 121 L (0.09 €/L)
H$_2$SO$_4$ or Na$_2$SO$_4$	https://www.sigmaaldrich.com/catalog/product/aldrich/339741?lang=it&region=IT https://www.sciencecompany.com/Sulfuric-Acid-Concentrated-32oz-P6550.aspx https://www.ladivinapiscina.it/it/ordini/?op=step1	194 €/100 g, 45 €/L or 32 €/8 kg [46 g/L] => 2.17 L (89.4 €/L)
		8 €/kg

CaCl	https://www.n2o3.com/en/catalogue/products/calcium-chloride_1832		[3 g/L] => 333 L (0.024 €/L)
Casein	https://www.iherb.com/pr/California-Gold-Nutrition-SPORT-Micellar-Casein-Protein-Unflavored-88-Protein-Slow-Absorption-Easy-to-Digest-Grade-A-Idaho-USA-Dairy-16-oz-454-g/71028		9.99 €/454 g [28 g/L] => 16 L (0.6 €/L)
Albumin or canned egg with (flavoured)	https://in.carbanio.com/chemical/albumin-9048-46-8-ch-xrgnemp or https://www.google.com/shopping/product/17038955685490733523?q=albume&sa=X&ved=0ahUKEwiarKv3_sPvAhVjwAIHHcfnAjUQkjAICCgA		100 €/10 g, 1.8 €/140 g [7 g/L] => 1.4 L (71.4 €/L)
Lactose	https://www.dailyvita.com/default/now-foods-lactose-powder-1-lb.html		7.49 €/2.2 kg [10 g/L] => 220 L (0.034 €/L)
Tot			≈ 70 €/kg

N.B. Fertilizer 3 (N 5 %, K_2O 16 %, P_2O_5 8 %, SO_3 25 %, MgO 3 %, Fe 0.1 %, B 0.02 %, Zn 0.01 %, 20 €/5 kg https://www.gogoverde.it/concime-universale-granulare-zapi-super-blu-8kg.html

5. Alveolar milk

Animal's milk-producing may be used for isolating mammary glands alveolar epithelium, grown and developed *in vitro,* for being transplanted back on the donor at the end of the growth culture in just couple of weeks, quicky recovering the beast.

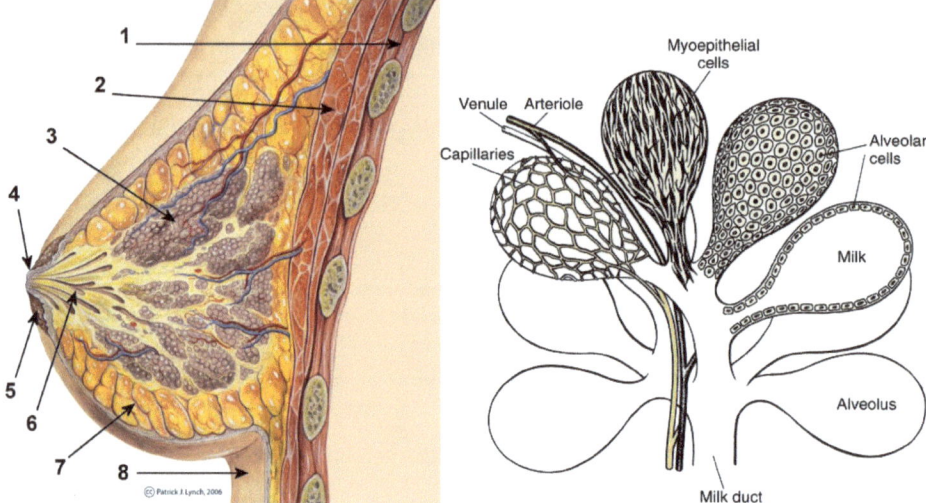

Figure 9 Human burst: 1. Chest walls, 2. *pectoralis major* muscle, 3. lobules (right), 4. nipple, 5. areola, 6. milk ducts, 8. skin (left).

Mammary alveolar epithelial cells, surrounded by myoepithelial cells, produce milk during **lactogenesis** in two stages: I. prepartum milk synthesis: sodium chloride, casein, lactose, lactalbumin, immunoglobulin, lactoferrin secretion into colostrum are regulated by progesterone; II. birth milk ejection: each of the teat cistern pair (V = 2 x 50 mL) is fulfilled of milk daily. **Mammary Stem Cell (MaSC)** traditionally can generate both the ductal and lobular structures of the mammary gland. Human luminal-restricted progenitor (EpCAMhiCD49f+MUC1+) (DOI: 10.12688/f1000research.18696.2). **Mammary** parenchyma tissue pieces present **primary cells** isolated digesting with trypsin-EDTA, filtering and centrifuging, seeded in Petri Dish containing antibiotics such as insulin, hydrocortisone, steroids and trypsin, and incubated at T = 37 °C, RH = 95 % and 5 % CO_2 (DOI: 10.1007/s11626-013-9711-4). **Bovine Mammary Epithelial Cells** (BMEC = 500 k CFU at t = 1 week) can secern casein on a plastic substrate thanks to plasmid gene transfer (DOI: 10.1271/bbb.59.59).

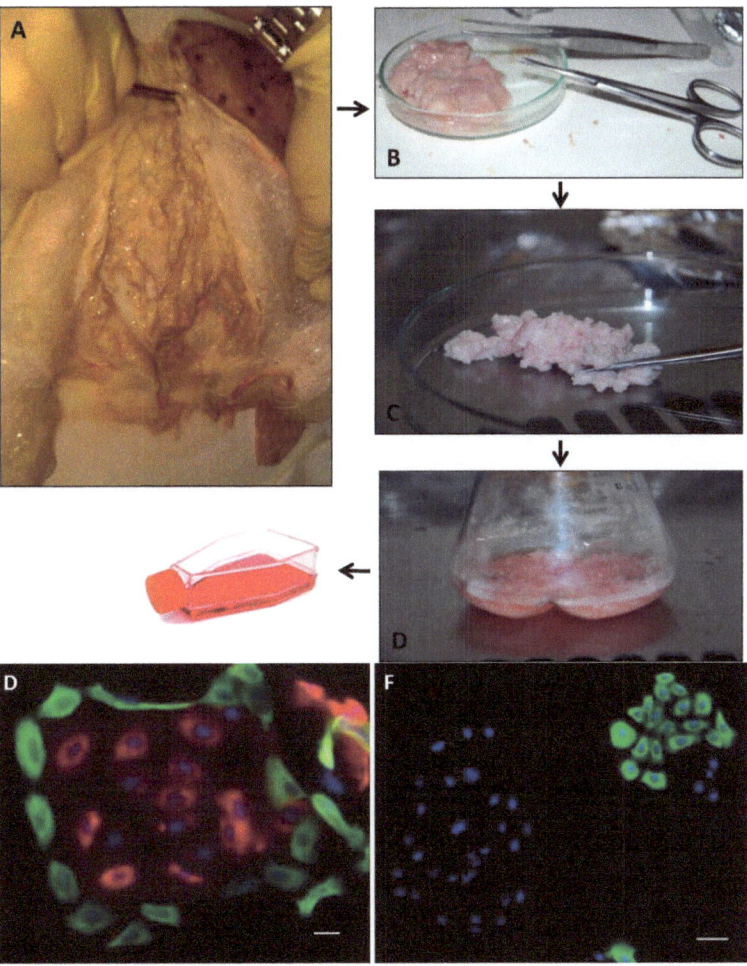

Figure 10 Mammary glands tissue (m = 100 g), wiped in ethanol 70 % (C_2H_5OH), can be dissected, minced, and digested with collagenase and hyaluronidase (400 U/mL) catalysed by sulfonic acid [R-S(=O)$_2$-OH] at T = 37 °C, filtering for isolating the primary goat Mammary Cell (pgMCs) centrifuging at v = 1200 rpm got t = 5 min, seeded on Petri Dish or suspended at ratio 1:5 in a growth medium with lactation hormones and trypsin them or storing in (90 % FBS 10 5 DMSO) (A-D top). Myoepithelial, immunostained with fluorescent, green cytological KeRaTin 14 (KRT14), encircling luminal epithelial cells with red KRT18 (D), organizing with them in alveoli in which both cells nuclei are coloured with blue DAPI (F) (bottom) – Image source: DOI: 10.5772/intechopen.71853

'Metagenomics – A pandemic theoretical and practical review on environmental genomics for basic research and applied industrial biotechnology' © Copyright Antonio Silvestro, 2019

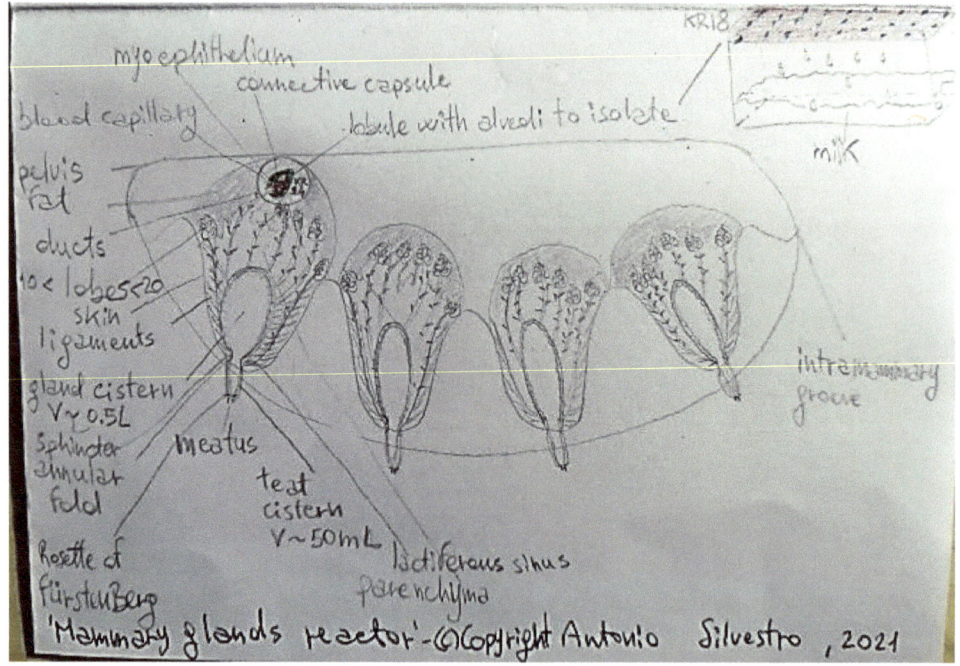

Figure 11 Milk-secernating **synth mammary glands** would derive as the natural both from mesoderm [white and red blood cells in blood vessels – thoracic artery and veins, plasma and nourishing fetal serum, collagen, laminin and lipids as connective tissue Extra Cellular Matrix (ECM) substitute] and ectoderm (cuboidal keratinocytes, myoepithelial cells, myocytes, supraclavicular and intercostal nerves), lymph nodes or antibiotics, oxytocin, vasopressin, parathyroid, estrogen, Growth Hormone (GH) and lactation hormones. Easily, may be structured an artificial **bioreactor** with an alveolar luminal epithelium (fluorescent stain with KR18 monoclonal antibody) as roof that would secern milk, even without the myosatellites contraction, falling according gravity into the storing container with a side basal tube to place it along an industrial productive series, or just used individually for retail scale and home. In this last case, the lactocytes roof may be removed and eventually replaced, before natural cellular senescence or in case of higher performance needed. Stem cell isolation from cow chunks, ribs, loin and round meat wastes such as connective, fat, followed by their differentiation in lactocyte would bring the total detachment from killing as method included in this dairy processing – Image source: *'Mammary glands reactor'* © Copyright Antonio Silvestro, 2021.

Bipotent, luminal and myoepithelial, Mammary Stem Cells (MaSCs) Sca-1 antigen is a Glycosyl PhosphatidylInositol (GPI)-anchored cell surface protein present in the lipid raft of the cell membrane and regulates many signalling events. Hormonal treatments like progesterone, estrogen and growth hormone, and by exogenous administration of xanthosine or inosine purine nucleotides (DOI: 10.1186/2049-1891-5-36). The Polycomb-Repressive Complex-1 (PRC1) gene Bmi1 (Liu et al. 2006; Pietersen et al. 2008) and PRC2 gene Ezh2 (Pal et al. 2013) influence mammary repopulating potential but play distinct roles in the developing gland. To gain insight into the global epigenetic signatures of human breast epithelial cells, CD44+ (enriched for putative progenitors) and CD24+ (luminal cell-enriched) populations were sorted, and their histone methylation and DNA methylation patterns were determined. H3K27me3 (frequently localized to K27 blocks in gene-poor domains) and DNA methylation patterns were distinct in the two subsets, suggesting that gene expression programs in the different cell types are controlled by epigenetic mechanisms (GENES & DEVELOPMENT 28:1143–1158 Published by Cold Spring Harbor Laboratory Press; ISSN 0890-9369/14; www.genesdev.org). Embryonic MaSC (fluorescent gen markers Axin2+, Notch1+, and PR/SET domain 1 Blimp1+) differentiate into unipotent MaSC luminal stem cells (Keratin8 K8+, E74-like factor 5 Elf5+, Prom1+, Notch1/3+ and Blimp1+), then in alveolar epithelial milk producing cells (DOI: 10.1155/2019/4247168).

Proliferation rate of 0.3 % and that of cell death of 0.56 % per day. Immortalized Bovine Mammary Epithelial cells (BME-UV1) and immortalized bovine Mammary Alveolar Cells (MAC-T) have the ability to synthetize caseins during cellular differentiation pathways implicated in lactation. Hydrocortisone, a-lactose, glutathione, bovine insulin, ovine prolactin, bovine holo-transferrin, hydrocortisone, L-ascorbic acid, penicillin, streptomycin, fungizone, and 0.25% trypsin/EDTA reagents. Cells were cultured in routine culture medium (mixture of DMEM/F-12, RPMI-1640, and NCTC 135 in proportions of 5:3:2 by volume) enriched with a-lactose = 0.1 %, glutathione = 1.2 mM, bovine insulin = 5 µg/mL, bovine holo-transferrin (5 µg/mL, hydrocortisone = 1 µg/mL, L-ascorbic acid = 10 µg/mL, 10 % (vol/vol) heat-inactivated fetal calf serum, and penicillin-streptomycin (50 IU/mL) in atmosphere of 5 % CO_2/ 95 % humidified air at T = 37 °C DOI: 10.1152/ajpcell.00261.2015).

Tissue culture dishes (ø = 6 cm) were coated with rat type 1 collagen solution diluted in 13PBS and incubated at T = 37 °C for t = 1 h. Goat mammary cells were mixed with ≈ 10^5 mouse 3T3 fibroblasts, which were previously treated with [mitomycin C] = 10 µg/mL for t = 2 h. Media used were human EpiCult-B (STEMCELL Technologies) supplemented with 5 % Fetal Bovine Serum (FBS), [hydrocortisone] = 10^6 M, [penicillin] = 100 U/mL, [streptomycin] = 100 µg/mL and Sf7 (Dulbecco modified Eagle medium [DMEM]/Ham F12 supplemented with 5 % FBS, [glutamine] = 2 mM, 0.1 % [w/v] bovine serum albumin, Epidermal Growth Factor (EGF) = 1 ng/mL, cholera toxin = 10 ng/mL, [insulin] = 1 µg/mL, [hydrocortisone] = 0.5 µg/mL (DOI: 10.1095/biolreprod.111.095489). Collagenous bone powder gel would support alveolar mammary epithelium cells and fibroblasts, isolated from meat wastes and Cu 13 % fungicide-treated, supplemented with [dry blood] = 7 g/L, or hydrophobic and polar DiChloroMethane = CH_2Cl_2 = 2 ng/mL https://www.lab-club.com/epages/mss.sf/?ObjectPath=/Shops/mss/Products/N-12471-1G&Currency=EUR&Locale=en_GB 18 €/g) and Solid Phase Extraction (SPE) for isolating hydrocortisone from saliva (DOI: 10.1365/s10337-009-1239-0) or urine 'salting out' employing ethyl acetate ($C_4H_8O_2$ 10 €/250 mL 10.1093/clinchem/13.10.855) https://www.ebay.it/itm/162174568292?chn=ps&mkevt=1&mkcid=28) or *Cyanobacteria* and isolation of penicillin from *Staphylococcus aureus*, *Penicillium notatum* and *Penicillium chrysogenum*.

6. Aristaeus cheese

Figure 12 The Greek God of cheese and bees to which as given the name to the 2135 Aristaeus asteroid in the 1862 Apollo group discovered in 1977 by E.F. Helin and S.J. Bus at Palomar Observatory (69239 Hermes, 2063 Bacchus, 4581 Asclepius, 3200 Phaeton, 14827 Hypnos, 4197 Morpheus) – Image source: François Joseph Bosio (1768–1845), (Musée du Louvre).

During the anaerobic process of (homo)**lactic acid fermentation** sugars are converted in pyruvate and then this in lactate by skeletal muscles and *Lactobacilliarophycae* LABs eubacteria (e.g. *Lactobacillus sp., Leuconostoc sp., Lactococcus sp.*) plant microbiome (hetero - phosphoketolase pathway generation also of ethanol and carbon dioxide, and bifidum pathway *Bifidobacterium bifidum* produce also acetate). More often the query is going to waste or has been developed a new taste is what will manifest in your mind sensing a new fermented product. Many of them emerging from the combination of physical conditions (T, RH, t) growing medium (e.g., refined sugars, milk, plants and fishes' wastes, etc.) and (micro)organism specie-breed-variety-strain(s) feeding on it.

Infused soft cheese (Latin: *caseus* 'to ferment, become sour'): rehydrate $V = 400$ mL infused milk powder with $V_{H20} = 600$ mL, boil, add 1 lemon juice and 1 tsp salt, refrigerate.

Figure 13 **Cottage** E226 (left) + E224 **vs Ricotta** (cooked twice, right) E330 **cheese** made from expired micro-filtered, pastoralized, homogenised and acidified skim milk ($V_m = 1$ L), heated for $t = 30$ min and finally separated into curd and serum two times. The lower acidity of the preservative used for making 'Ricotta' = ($m_r = 50$ g) formed lesser cheese and a 2nd serum ($m_s = 950$ g : 8.5 g proteins).

1) $V = 3$ L cow milk : $m = 250$ g soft cheese (1 Mozzarella), $V = 1$ L cow milk : $m = 250$ g ricotta. N.B. Volume *Bos taurus : Ovis aries-Capra aegagrus hircus's* milk $= 10$ L : 6 L;

2) Production of the ***starter culture***: washing, sanitizing, sterilizing *in situ* substrates, fermenting sourdough, centrifuge, harvesting, lyophilization, granulation, packaging and labelling. Starter screening in accordance with the acidifying, proteolytic, peptidolytic

activities, bacteriophage resistance, antagonistic, aromatizing, lipolytic, antioxidant assets and polysaccharides biosynthesis. The bacterial starter (1 kg : 1 L = m : V skim milk UHT) clotting the milk reduce its pH from 6.75 to 4.5, and calcium (Ca) and phosphate (P) move down into the whey. Mother culture made from skim cow milk last t = 1 month and can be used for frequent cheese making (each Δt = 2 days). Native starter can be made from cow milk, resembling Yoghurt, in which lipases esterified Short Fatty Acids (SFA) reacting with Dionysius alcohol makes the milk pinkish emanating a fruity smell, or sheep milk for buttery milk. Natural sour cream can be made just leaving the milk out of the fridge overnight in a dry and shaded place. <u>Mesophilic</u> (T_{max} = 40 °C) *Lactobacilus subs. lactis (Streptomyces lactis), L. subsp. cremoris (S. cremoris)*, [1 €/7 L] (e.g., Cheddar, Stilton, Gouda, Blue) [15 €]. <u>Thermophilic</u> (45 < T < 60 °C) *Streptococcus thermophyilus, Lactobacillus delbrueckii subsp. bulgaricus, Lactobacillus helveticus* (e.g., Mozzarella, Parmesan, Provolone, Romano, Swiss, Gruyere);

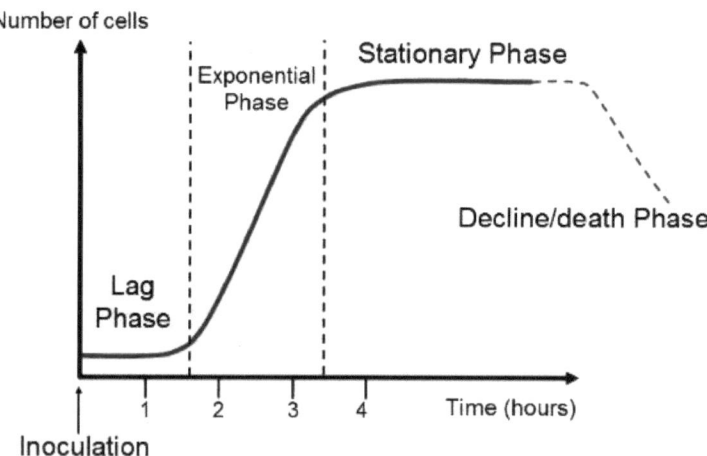

Figure 14 Colony Forming Unit (CFU) reaches the highest asymptotic point of the sigmoidal growth curve at 7 days and stay in the stationary phase for about another week, after which can store in the refrigerator at T = 4 °C. Just add 1 tbs of sodium chloride (NaCl) for improving its preservation avoiding contamination of non-natives microorganisms.

3) Acidifying **rennet**: microbial (*Rhizomucur pussillus, Rhizomucur miehei* isolated from *Cynara scolymus, C. cardunculus, Malvaceae spp., Ficus carica, Urtica dioica, Ananas comosus, Cucumis melo, Carica papaya* left-overs wrapped in micro-crystalline cellulose capsules with Mg-stearate as anti-clotting. 1 average rennet *tablet* (crushed and dissolved in VH_2O = 5 mL) [10 tabs about 6 €] : 18 L milk, 29 < T < 35 °C, t = 45 s (storage in fridge up to 3 years, N.B. animal rennet is characterized by a longer time of conservation). *Powder* 3 g/L (844 €/kg), *liquid* 1 tsp : 20 drops : 1 mL : 1 Tab Dilution 1 : 3 (540 €/L). 0.2 mL = 4 drops : 1 kg milk at T = 30 °C for t = 35 s, coagulation for t = 15 min and setting t = 30 min. International Milk Clotting Units (IMCU) 1 IMCU : 23 L milk (e.g. 60 IMUC 10 mL rennet/L milk). Single strength rennet: US 1:15000 (280 – 300 IMCU), Europe 1 : 10000 (200 IMCU). Meito *Zygomycota Rhizomucor pusillus, R. miehi* 7500 IMCU/g (5 g Rennet : 2 L milk). Chymax Christian Hansen, recombinant *E. coli* K12 2080 IMCU/g 0.11 mL, C.H. max liquid 30 mL : 1000 L milk. Fungi like *Acomycota Saccharomyces lactis, Aspergillus*

niger containing aspartic proteases as milk clotting. Animal Abomasum 4th stomach ruminant calf (95 % chymosin, 5 % pepsin). Human saliva (95 % pepsin) proteolysis inducer, makes the milk smell acid and cheese. *Rennet clotting test* via Dionysius ethanol 70 % precipitation in V = 5 mL. Casein A, B and K in micelles acidified with citric acid [m_{E330} = 30 g (5 €/kg)] or acetic acid (E 260) make Ricotta and Cottage cheese, respectively. Some orchard trees and plants such as *Vitis vinifera, Citrus x aurantium* and *Malus sylvestris* fruits can be used for soft cheese production due to their acid content [e.g., apples acids: malic (E 996), acetic, citric (8 mg) and ascorbic (E 300)];

Figure 15 Microbial rennet (*Rhizomucur spp.*) isolated from artichokes (*Cynara scolymus*) growing at 24 < T < 55 °C and 50 < RH < 80 %, visualized with Hamlet Digital Microscope at 50 X – Image source: © Copyright Antonio Silvestro, 2021.

4) *Sodium chloride* (m_{NaCl} = 30 g);
5) **Lactose** pills in the growth medium, broken-down by the lactase catalyses secreted by microorganism in the milk of the *Lactobacilliarophycae* family fermenting the substrate into glucose and galactose;
6) Herbs;
7) Hardening ([$CaCl_2$] = 0.3 g/L, 2.5 €/kg) for setting and make the cheese firm for cutting (not used in soft hand-cutted Mozzarella), can be made reacting NaCl with $CaCO_3$ from egg shells (isolation protocol in *'Culture and Cultures'* © Copyright Antonio Silvestro, 2020 Kindle eBook 3.89 € https://www.amazon.com/gp/product/B08DL76M5T/ref=dbs_a_def_rwt_hsch_vapi_tkin_p2_i2).

Materials:
1. Moulds
2. Thermometer (50 < T < 300 °C) [7 €]
3. Plastic plate
4. pHmeter
5. Cotton cloth bleached
6. Two steal bowls
7. Tarated cylinder
8. Magnetic agitator
9. Scale

10. Cheese cloth

Yeasts makes bubbles in some cheeses like Emmental and Asiago via carbon dioxide (CO_2) production through Dionysus *ethanol fermentation* of the microbiome.

Storage: plastic envelopes wavy out, porous in.

Among the most delicious Italian hard ovine cheese, the **Pecorino**, for which a m = 1.5 kg would be produced heating pasteurized sheep milk V_{milk} = 10 L (optional extra fermenting starter *Lactobacilliarophycae* bacteria) in a steel bowl at T = 28 °C, adding V_{rennet} 1.5 mL (strength 1:10^5) stirring leaving it cuddling for t = 20 min, heat again at higher temperature T_2 = 37 °C, filtrate the granular cuddle from the serum, leave it for t_2 = 12 h twisting upside down each half hour (6 times), store in fridge at T_3 = 4 °C for t_3 = 1 day, finally, dry superficial or brine salt (NaCl) for t_4 = 36 h.

Cheese	Features	Origin	Lactic animal
Keen	Quality Cheddar	Davon - UK	cow (*Bos taurus*)
Montgomery	Cheddar	UK	cow
Taw Taste Valley	Cheddar	UK	cow
Cham wood	Cheddar paprika	UK	cow
Manchego	hard, strong, grating	Spain	ewe (*Oris aries*)
Grana Padano	15 months ripening	Italy	cow
Parmiggiano Reggiano	25-36 months ripening, hard, strong, grating, insulation via acidifying anaerobia *Lactobacilliarophycae* (pH < 4, ethanol-> acetic acid) forbidden for storing the cattle foraging	Emilia Romagna - Italy	cow
Pecorino	hard, strong, grating	Sardegna - Italy	ewe
Rachel	unpasteurized, fruity	Cornwall	goat (*Capra aegagrus*)
Scarpham Brie	creamy, crumbly, UHT, vegetal rennet	UK	goat
Scarpham Rustic		UK	goat
Emmentaler	*Streptococcus thermophilus*, *Propionibacterium freudenreichii*, *Lactobacillus helveticus*	Switzerland	cow
Gruyére		Switzerland	cow
May field	Emmental	UK	cow
Cornish Yarg	nettle (*Urtica dioica or U. monoica*)	Cornwall - UK	cow
Garlic Yarg	*Allium sativum*	Cornwall - UK	cow
Gorgonzola	green mold	Lombardia - Italy	cow
Devon smoke		Devon - UK	cow
Dorsel Vinnie blue		UK	cow
Osau Isayy		France	cow
Feta	brine	Greece	cow, ewe
Ricotta	curdle by-product, spreadable	Italy	cow, ewe
Cottage Cheese	curdle by-product, spreadable	UK	cow, ewe
Cheivre	curdle by-product, spreadable	France	cow, ewe
Mozzarella	hand-cutted, stretched, soft	Campania - Italy	buffalo (*Bubalus bubalis*)

'Metagenomics – A pandemic theoretical and practical review on environmental genomics for basic research and applied industrial biotechnology' © Copyright Antonio Silvestro, 2019

Trehill	garlic, chives		cow
Quickie's	ripen six months		goat
Quickie's vintage			goat
Creamy goat	acid		goat
Brigante	mild	Sardegna - Italy	ewe
Devon Blue	greenish/bluish mould (*Pennicillum roqueforti, P. glaucum*)	Davon UK	cow
Roquefort	greenish/bluish mould (*Pennicillum roqueforti, P. glaucum*)	France	cow
Stilton	greenish/bluish mould (*Pennicillum roqueforti, P. glaucum*)	UK	cow
Camembert	white chalky crust with soft, runny and sticky core (*Pennicilum camemberti*)	France	cow
Harbonne	creamy, blue mould	UK	cow
Reicnester red Westcombe	orange dyeAnatto (*Bixa orellana*)	UK	cow
Bernleigh blue		UK	cow
Ntufu		Asturias - Spain	cow
Asiago	Mild, holed	Italy	cow
Montagrolo		Germany	ewe
Cabrales	mould	Cantabria - Spain	cow
Delice de Bourbogne		France	cow
Epoisse de Bourbogne	brine adding in dissolving common salt (NaCl) in water/bree/wine with spices.	France	cow
Appenzeller	pungent brine, *Brevibacterium lineus*	Switzerland	cow
Black bombei	wax rind	Cornwall (UK)	cow
Vignette			cow
Abergoveny	soft, spread	UK	goat
Brie	smear-ripened, brine, rindless	France	cow
Munster	semi-soft, high moisture	Germany	cow

Do not forget that if you have not your own green field, raise cattle would not be affordable and healthy despite you consume it moderately remember the high ration of milk secreted by the mums for making it, and that during World Wars, cheese was used by soldier for easy carry a high fat content food. Again, spread the hard onto pasta spaghetti, melt the mild into delicious sandwiches and the soft with veggies like the typical caprese with tomatoes of the Mediterranean cousin, world heritage.

Table 1 Fermented food feedstock and starter microbial culture.

Fermented food feedstock	Starter microbial culture
Flour	Yeast (*Saccharomyces cerevisiae, S. carlsbergensis, S. pastorianus, S. exiguus, Brettanomyces bruxellenensis. B. anomalus, Saccharomyces cerevisiae, S. bayanus, Kloekera spp.*), LABs,
Cereals	*Candida spp.*, LAB, yeast, molds, *Pennicillum camemberti, P. roqueforti*
Grape	*Geothrichum candidum*
Milk	LAB, *Firmicutes/Lactobacillus* and *Staphylococcus*, yeast and mold.
Meat	LAB
Fermented veggies (E.g. *Cucumis sativus, Olea europeae*, etc.)	SCOBY (Symbiotic Culture of Bacteria and Yeast)
Soy sauce (*Glycine max*)	
Sauerkraut (*Brassica oleracea*)	
Milk Kefir, Tibicos, Ginger (*Zingiber officinale*). Beer, Kombucha tea	

Starter can be classified according to microbial complexity, T (°C, K, F) *optimum*, lyophilization and other storages processes. Morpho-physiology, *modus operandi* (directly or progressively in a boiler).

Wild ("natural") Starters (WS) can be made with the "*Back-slopping*" technique using an *inoculum* from the same fermented food or an antecedent well-done preparation (Natural Whey Culture: thermophile (40-45°C) Grana, Provolone, Pecorino Romano, mesophile (20°-30°C) Water buffalo Mozzarella, milk cream; lactic starter thermophile Crescenza, Asiago, mesophile Caprini, Pecorini, milk cream or sourdough).

- Starter *Lactobacillariaceae* - SLAB (mesophile): *Firmicutes/Streptococcus thermophilus* and *Lactococcus lactis;*

- UnSuitable *Lactobacillariaceae* - USLAB (thermophile): *Actinobacteria/Micrococcus, Firmicutes/Enterococcus, Pedicoccus, Leuconostoc.*

Screened starter culture, Generally Recognized As Safe (GRAS), in accordance with the acidifying, proteolytic, peptidolytic activities, bacteriophage resistance, antagonistic, aromatizing, lipolytic, antioxidant assets and polysaccharides biosynthesis.
Production starter: washing, sanitizing, sterilizing *in situ* substrates, fermenting sourdough, centrifuge, harvesting, lyophilisation, granulation, packaging, labelling.

Molecular monitoring methods:
species-specific *in situ* or on isolates: 16S rRNA gene, Internal Transcribed Spacer (ITS), 510 PCR-DGGE, Terminal Restriction Fragment Length Polymorphism (TRFLP or sometimes T-RFLP);
strain-specific just on isolates: Random Amplified Polymorphism DNA (RAPD-PCR), Restriction Endonuclease Assay - Pulse Field Gel Electrophoresis (REA-PFGE).

'Timeless fermentations' © Copyright Antonio Silvestro, 2019

Figure 16 *'Aristaeus candle'* made with cheese rind wax recycled, melting it in a metal pot during couple of minutes adding some Athena olive (*Olea europaea*) oil (3:1), cooling and solidifying in about t = 30 min at room temperature - Image source: (C) Copyright Antonio Silvestro, 2021.

7. Cheesecake

- Crust: 180 g crumble/crackers, 30 g sugar, 45 g infused butter, ¼ tsp salt
- Filling: 250 g infused soft cheese, 1 egg, 35 g sugar, 1 tsp vanilla, ¼ lemon

Preheat oven T = 180 °C, fill the tray of water, wrap the bowl with aluminium foil out and parchment buttered paper in, bake for t = 30 min at T = 180 °C, more t = 30 min at T = 150 °C, cool in oven t = 1 h, refrigerate t = 6 h, m = 40 g berries infused syrup.

8. Yogurt

Yogurt is produced by pasteurizing Ops milk (V = 150 mL) at temperature T = 85 °C, cooling it down to temperature $37.5 < T_{opt} < 42$ °C overnight for t = 8 h via bacterial lactic fermentation of *Lactobacillus delbrueckii subsp. bulgaricus* and *Streptococcus thermophilus* 10^6 Colony Forming Unit (CFU)/mL converting lactose in the milk into lactic acid. It is mainly characterized by the following nutritional values: 81 % H_2O, protein 9 g, carbohydrates 4 g, fat 5 g, riboflavin (B_2 23 %), vitamin (B_{12} 31 %) and phosphorous (P 19 %).

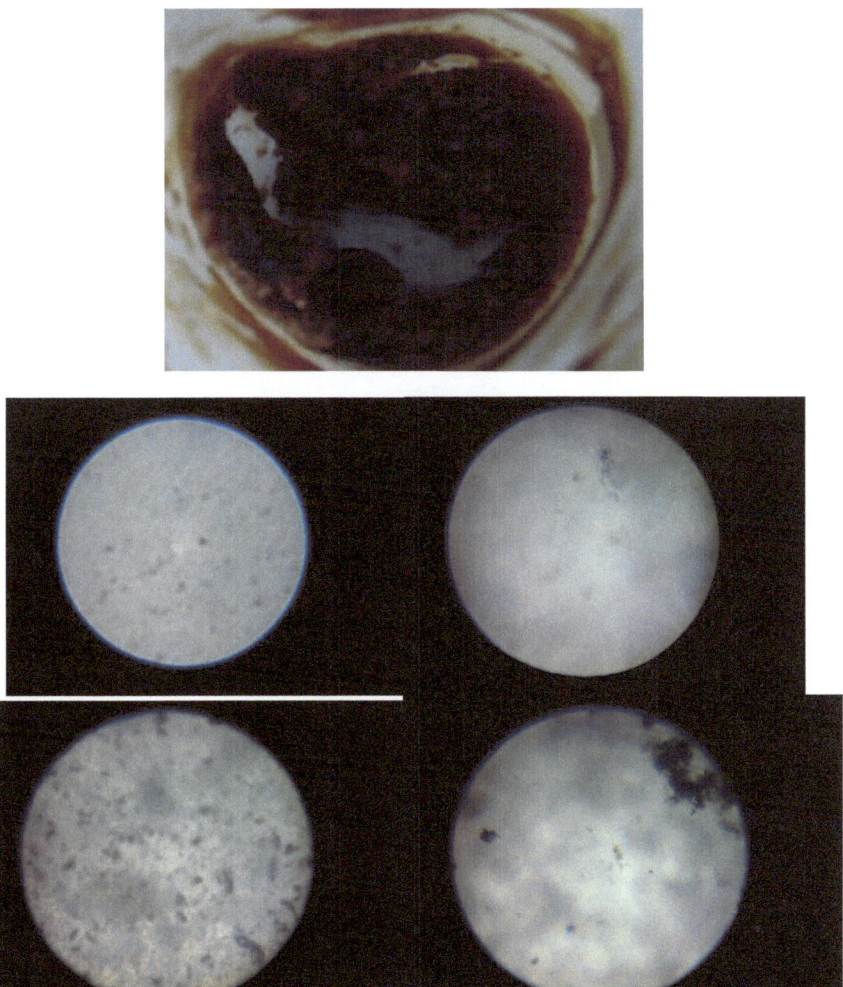

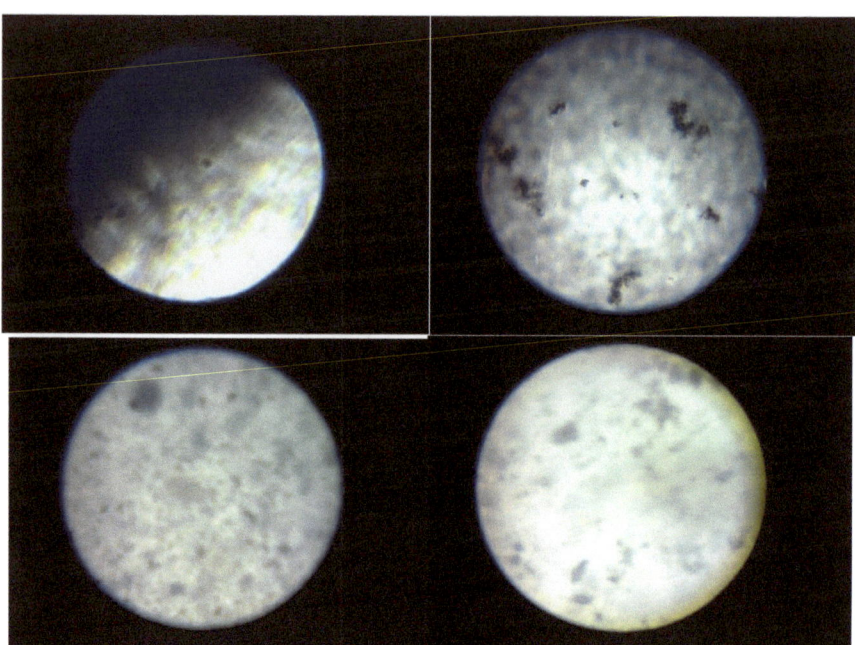

Figure 17 Brownish yogurt stored for t = 3 years in the fridge at T = 4 °C in which spoiler coccoidal orangish molds are growing contaminating the native microbiome composed of *Lactobacilliarophycae Lactobacillus bulgaricus* and *Streptomyces thermophilus* magnified at 50, 100, 200 and 400 X with Hamlet Digital Microscope – Image Source: © Copyright Antonio Silvestro, 2020.

9. Ice cream

Ice creams are frozen desserts constitute mainly milk to which are added whisked cream and sweeteners such as sugar or stevia, spices like cocoa and vanilla mainly for covering the buttery flavour due to its main components. Its foam as to be preserved at temperature below T = 2 °C.

Ethanolic fermentation - Dionysus/Bacchus - the spirit of enology and zymology

Rebirth in your liver Pluto in Scorpius and thyme Sun in Leo, is against alcoholic beverages of Saturn in Capricorn, for which is needed to drink them eating calcinated food Mars in Aries like bread and meat.

Yeasts unicellular *Eukarya* (1.500 species, *Fungi Ascomycota + Basidiomycota*, $\varnothing \approx 3$ µm, E = 1.4 kJ, $m_{carbohydrates}$ = 41 g, $m_{protein}$ = 40 g, m_{fat} = 7.6 g per 100 g, vitamins: B1, B2, and B9, elements: P and Zn) evolved from rebel multicellular that decided along the evolution line to disassemble themselves from one only neuronal control centre into many individuals, chemoorganotrophs processing the source of energy in need to their ancestors for being free, precisely, converting carbohydrates $[C(H_2O)_n]$ into carbon dioxide (CO_2), that at $[CO_{2(g)}] \geq 70$ k ppm may cause

'Timeless fermentations' © Copyright Antonio Silvestro, 2019

suffocation, preventable changing the phase into liquid raising the pressure at $p \geq 5$ atm, and ethanol (C_2H_5OH). As yeast would have been ejected from the originary pluricellular, as they were triggering a kind of parasitic relation with them. Hence, pushed away from their body, generally, budding out reproducing asexually, conjugating the haploid cells into diploid spore- forming. Top- *Saccharomyces cerevisiae,* form superficial foam on the wort, worth to save sugary waste substrate (ale), and bottom-fermenting yeasts *S. carlsbergensis* (lager), can be obligate or facultative aerobes, but not obligate anaerobes, feeding for example on sucrose disaccharide ($C_{12}H_{22}O_{11}$) extracted form *Saccharum officinarum* (3 €/20 seeds https://www.wish.com), optionally clarified using calcium phosphate or carbonate (CPO_4^{3-}, $CaCO_3^{2-}$), but also on organic fibre bagasse wastes of this plants and the other of its genre generating ethanol and carbon dioxide.

Syrup: hot plant infusion (V = 500 mL), sugar (combination sucrose, glucose and maltodextrin, m = 125 g), and food colouring.

Ethanolic fermentation: syrup medium (V_{H2O} = 0.5 L), $m_{Glucose}$ ($C_6H_{12}O_6$ = 125 g) or equivalent starch from boiling water and yeast cube ($m_{Saccharomyces}$ = 7.5 g). Just Cancer dissolving fruit peel in water may let you generate a fermented drink, but adding extra yeast starter (e.g., *S. cerevisiae*) and refined sugar sucrose from sugarcane (*S. officinarum*) will speed up the ethanol production. Pectinases enzymes may catalyse the hydrolysis of the fruits exocarps. Furthermore, if wishing to make inedible bioethanol fuel, is possible to add hydrolytic sulfuric acid, balancing the acidity due to it with the strong base sodium hydroxide, for increasing even more the velocity of fermentation.

Cellulase (endoglucanases, exoglycanases and cellobioses) and amylase enzyme for breaking down structural cellulose and storage starch freely glucose reagent of ethanolic fermentation, may be isolated from organic biomass Fungi (*Ascomycota* and *Basidiomycota*), cellulolytic bacteria (e.g., *Clostridia* and *Ruminococcus*) growth in a medium made of 1 % peptone (animal digest), 1 % CarboxyMethylCellulose (CMC), 0.2 % potash hydrogen phosphate (K_2HPO_4), 1 % agar, 0.03 % magnesium sulphate ($MgSO_4$) x 7 H_2O, 0.25 % ammonium disulphate ($NH_4)2SO_4$ and 0.2 % gelatine at pH = 7 for t = 2 days of incubation at T = 30 °C. (DOI: 10.5505/tjb.2012.09709).

Both yeast (*Saccharomyces cerevisiae, Pichia stipites, Pachysolen tannophilus, Candida tropicalis,* and *Candida shehatae,*) and bacteria (*Zymomonas mobilis*) can ferment organic lignocellulosic biomass (40 % cellulose, 30 % hemicellulose and lignin 20 %) for generating bioethanol fuel (5.5 + 8 € shipping https://www.manomano.it/p/bioetanolo-combustibile-2-4-6-8-12-24-litri-per-caminetto-liquido-camino-stufa-5305417#/) from glucose via glycolysis ($C_6H_{12}O_6 \rightarrow 2\ C_2H_5OH + 2\ CO_2$ + heat) or parallel pentose phosphate pathway (Fermentation 2018, 4, 16; DOI: 10.3390/fermentation4010016). Substrate pre-treatment can be done with fungal digestion (*Aspergillus terreus, Aspergillus awamori, Trichoderma reesei, Trichoderma viridae, Ceriporiopsis subvermispora, Echinodontium taxodii, Irpex lacteus, Phanerochaetechrysosporium, Pleutus ostreatus, Pycnoporus cinnabarinus, Pycnoporus sanguineus* and *Fusarium concolor*) (DOI: 10.21741/9781644900116-12), or via kraft pulping sulphate process with which the biomass substrate is broken down with sodium hydroxide (NaOH) and sodium sulphide (Na_2S).

Cytoplasmatic carbs (m = 60 g ≈ 2 tbsp) can be extracted from biomass using sulfuric acid (96 % V_{H2SO4} = 200 mL ≈ 1 PP glass, 8 €/L https://www.letslab.it/acido-solforico-epr.lab) in t = 1 day, this could be isolated from Zeus/Jupiter acid rains or, perhaps, its esterified form from Aphrodite/Venus in Libra urines, and then from each gram could be isolated enough glucose for fermenting into V_{C2H5OH} = 0.25 mL in t = 1 week ($V_{C2H5OH\ tot}$ = 15 mL).

'Timeless fermentations' © Copyright Antonio Silvestro, 2019

Dionysus ethanolic fermentation in Tetra Packs at room temperature T = 25 °C for t_1 = 2 days, followed by Gemini refrigeration, feeding weekly (t_2 = 7 days) let doubling at room temperature T = 25 °C for t_3 = 12 h, then saving one half and use the other as feeder for starting the next batch culture multiplied by Aquarius zodiac:

1. Lukewarm water V_{H2O} = 100 mL;
2. Yeast (*Saccharum cerevisiae*) m = 50 g;
3. Sucrose ($m_{C12H22O11}$ = 250 g), pasta/rice/potatoes boiling water whitish jelly starch $[V(C_6H_{10}O_5)_n$ = 25 mL] for making beverage, or flour (m = 1 kg) for baking bread;

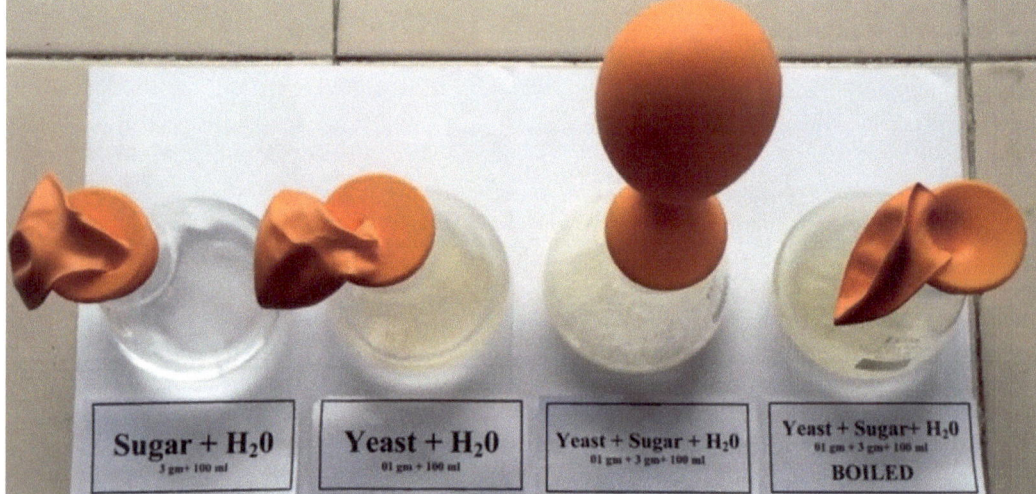

Figure 18 Ethanolic fermentation

4. V_{H2O} = 1 L.

Yeast (*Saccharomyces cerevisiae*) can be used as feed for *Tubitrex aceti, Panerellus redivivus, Gallus gallus, Oryctolagus cuniculus* and *Carassius auratus*. It lasts 10 days, for which is advisable to supply itself weekly, into a fermenter, with sugars.

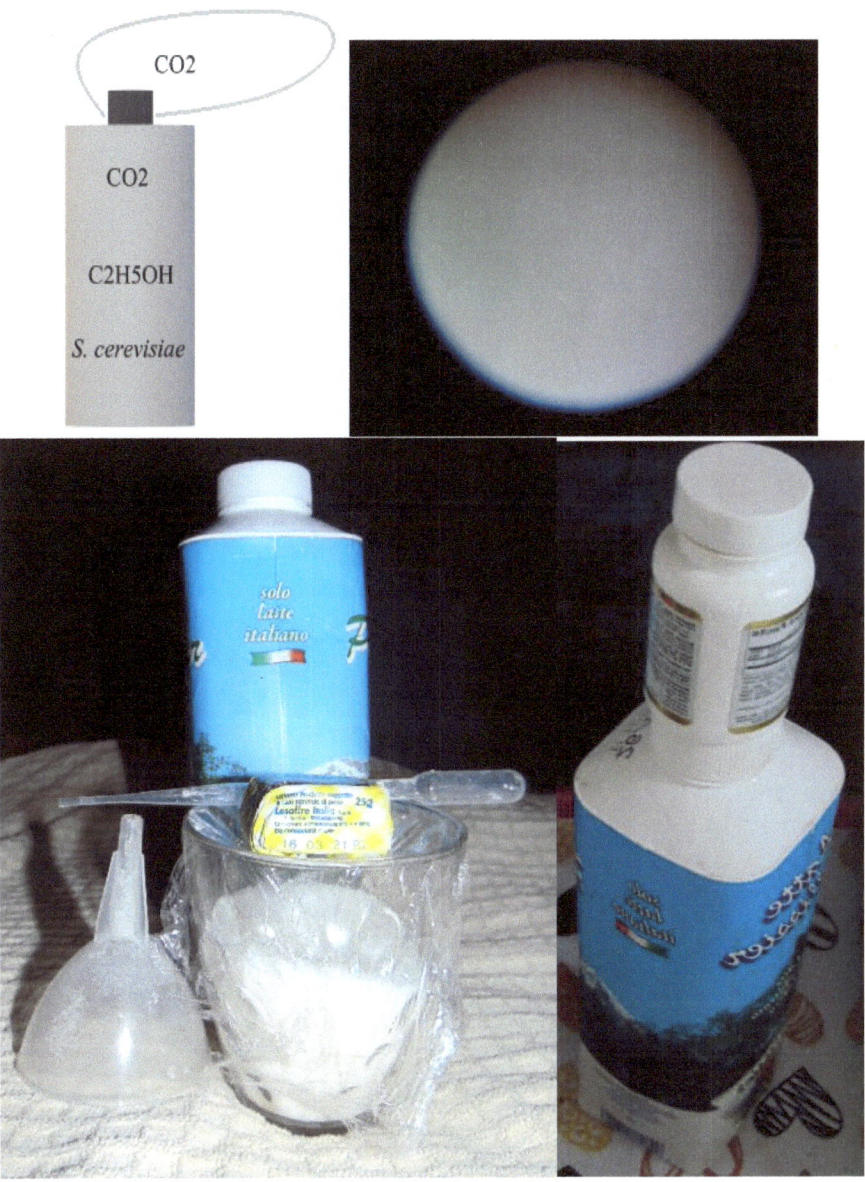

'Timeless fermentations' © Copyright Antonio Silvestro, 2019

Figure 19 **Yeast** (lyophile parallelepiped *Saccharomyces cerevisiae* $20 \cdot 10^9$ cells/g, m = 75 g, 3 x 25 g : 0.3 € + 5.5 € shipping https://www.laspesa24.com/lievito-di-birra-fresco/36563-lievito-fresco-mignon-g25x2-lievitalia--8009511010418.html, Colony Forming Units (CFU) dissolved in the alcoholic solution account =

$4 \cdot 10^5$ CFU/mL, duplication time ≈ 2 h, visualized with Hamlet Digital Microscope at 50 X) grown and developed in in white ThermoPlastic PolyEThylene mini-bottle (V_{PET} = 75 mL), single, double and tetra Tetra Pack, PolyEThylene (PET) bottles (V_{PET} = 1 and 2 L) and glass fermenters each with air-lock, and one on magnetic stirrer. Sucrose (m = 200 g) dissolved in water (V = 1 L), is inverted into fructose and glucose, this last undergoing **ethanolic fermentation** expiring asymptotically into carbon dioxide ($m_{CO2(g)}$ ≈ 200 g in just t = 12 h, ρ = 2 g/L, critical point at T = 29 °C and p = 100 atm), generating up to about **12 %** of **ethanol** (C_2H_5OH, E = 21 MJ/L), at optimum Temperature 25 < T < 33 °C (mesophilic) and Relative Humidity 75 < RH < 95 % for t = 2 weeks.

Transparent PolyEThylene (V_{PET} = 1 L) fermenters with yeast (*S. cerevisiae*) fed with 'mother sourdough' starchy polysaccharide (300 < x < 1000 glucose) **flour** (left - m = 50 g, V = 1 L, ρ = 50 g/L) or **sucrose** disaccharide (glucose + fructose) (right - m = 400 g, V_{PET} = 2 L, ρ = 200 g/L), characterized by a mass ratio of 1/4, for slow and fast fermentation comparison assay, respectively, with theoretical velocity difference of a factor 3, depending on the diverse energy linking the monosaccharides of different stereoisomery on the anomeric hemiacetal(ketal) carbon C1 involved in α-1-β-2 C-glycosidic covalent linkage of the sucrose, hydrolysed by invertases, while, α1-4 in linear amylose and α1-6 in branching amylopectin C-glycosidic bonds in the starch, broken down by amylase secreted by the salivary glands while biting warm just baked delicious bread. Sourdough supernatant (V = 0.5 L) and Ceres's flour (m = 500 g) in a PE bottle (V = 1 L) levitating from Capricorn 11:00 to Aries 13:00 before being baked in hoven. If caramelized molasses due to uncompleted ethanol fermentation happen, use it again to feed your yeast.

Carbon dioxide ($CO_{2(g)}$) inspirable with a syringe, may be changed into liquid ($CO_{2(l)}$) state reducing temperature 22 < T < 29 ° C and pressure p = 10 atm for a safer utilization, may be let react with Sun hydrogen [$H_{2(g)}$ E = 10 MJ/L)] generated with electrolysis (ΔV = 1.2 V, 120 < p < 200 atm) of Jupiter rain-water deionized via Virgo fractional freezing (H_2O) at cathode (+), forming Capricorn alcohol (e.g. methanol), Taurus petrol fuel (hydrocarbon C4–C12, ρ = 75 g/L, E = 47 MJ/kg, E = 36 MJ/L, e.g. methane CH_4 + 2 H_2O) via thermal catalysts, for boosting photosynthetic CO_2-fixation into carbohydrates (e.g. formaldehyde-methanal) nebulizing Ceres plant's leaves or submerging Uranus phytoplankton into it – Image source: © Copyright Antonio Silvestro, 2021.

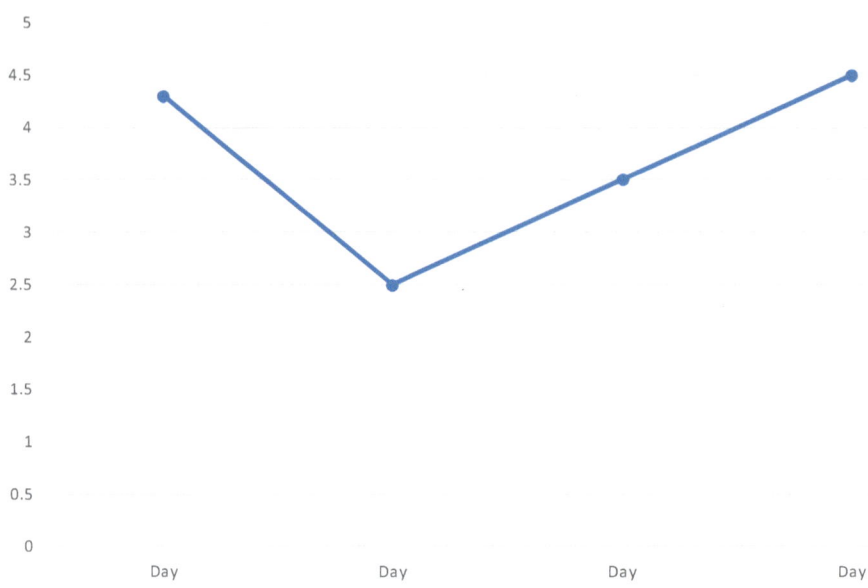

The Holy Grail, the magic potion drunk by Jesus of Nazareth during the last supper, should have been close to an archaic beverage produced with barley grains similar to the beer, along which many legends have been told and written among which that concealing the King of Arthur Round Table. Actually, **beer** is made using malt powder extract or self-made malted barley sprouted [*Hordeum vulgare* $t_{soaking}$ = 2 days, $t_{germination}$ = 5 days, ~ 100 sprouts/1 L beer (t_{1a} = 1 week old), kilning of the germinated sprouts directly by the syrup not cooled down, with hot hair flow or in incubator at T = 37.5 °C for t = 1 day, room temperature T = 20 °C for t_2 = 1 week, and finally, remove the rootlets. After 2 weeks curing the hulled rootless sprouts, (they may taste sweet, pungent on the tip of the tongue and slightly frizzy), are milled into flour determining the colour of the beer (e.g., pale amber Lager and Pilsner, dark stout Guinness). To them may be added other starch sources such as wheat (*Triticum aestivum*), corn (*Zea mays*), rice (*Oryza sativa*), oats (*Avena sativa*), rye (*Segale cereale*), millet (*Panicum mileaceum*), gluten-free sorghum (*Sorghum bicolor*), cassava (*Manihot esculenta*), agave (*Agave* spp.), thoghter mashed in hot water (93 %) at 50 < T < 70 °C ($T_{saccharifications}$ = 65 °C), for t = 2 h into tun for breaking down amylose and easily amylopectin of the starch carbohydrates into polymeric dextrin, later into tri-saccharides, di- (e.g. maltose) and monosaccharide glucose. The cooled syrup (V_{wort} = 500 mL) filtered from the solid mash by-product (lautering) and mixed to bittering and flavouring hops (*Humulus lupulus*) boiled, in electroconductive and fungicide copper (Cu) bowls for t = 1 h, whirlpooled, cooled in capacitative heat exchanger (T_1 = 95 °C <-> T_2 = 10 °C), to which is added lyophilized yeast starer [$m_{starter}$ = 2 g in $V_{lukewarm\ water}$ = 10 mL at T ~ 37.5 °C stirring for t = 5 min, *Saccharomyces cerevisiae* at T = 20 °C, *S. pastorianus* at T_1 = 10 °C (fermentation stage) and 0 < T_2 < 4 °C (lagering phase in Gemini fridge), *Brettanomyces bruxellensis*, *B. Brettanomyces lambicus* and *Torulaspora delbrueckii*] fermenting glucose ($C_6H_{12}O_6$) into ethanol (C_2H_5OH, 4 < ABV < 40 %)

'Timeless fermentations' © Copyright Antonio Silvestro, 2019

and carbon dioxide (CO_2) [yeast test: 1 tsp sugar -> foam], conditioned in barrel with air lock (2 weeks < $t_{conditioning}$ < 3 months), carbonated for bottling adding the priming sugars ($m_{glucose}$ = 25 g, $m_{sucrose}$ = 33 g or $m_{maltodextrin}$ = 40 g), chemically acidifying the solution using *Rutaceae* such as *Citrus limoni*'s citric acid or *Citrus sinensis*'s ascorbic acid (≈ Vitamin C) or lactic fermenting bacterial *Lactobacilliarophyceae*, and finally, 1 tsp kitchen salt (NaCl) (Beer make kit 35 € https://www.troppotogo.it/kit-per-birra-pilsner-fai-da-te?cpkey=v7nNtHd1rfm7BVlfdlAUbxqqPMY5mOEkX2CDBE9WV_KTzjBUWWHGq-sHL5torzNn&gclid=Cj0KCQjw5eX7BRDQARIsAMhYLP_TxKqKJVDjrmOZwIQHt_yfQwjTA4B9PXZ6rwIWnLKt-w4h0wpUN4EaAnR1EALw_wcB). The Holy Grail, should have been made with the blessed Ceres grains, currently, produced worldwide as many types and brands of beer, for example, the Arthur Guinness Irish beer, that may let you remind the wife of the King, Genevieve, that cheated him with Sir Lancelot du Lac, leaving the knight Sir Bedivere throw the Excalibur sword ordered by the ashamed knowing the fact.

Figure 20 'Bacchus' by Caravaggio, oil on canvas (1596–97), in the Uffizi Gallery, Florence, Italy.

Ambrosia (Ancient Greek: ἀμβροσία, 'immortality') the ingested conferring oblivion of the past, immortality in the present and longevity for the future.

Wine is an acholic beverage characterized by a variable fungicide and bactericide ethanol volume (5-15 %), produced from fermented (*Saccharomyces spp.*) grapes (*Vitis vinifera*) or other fruits such as *Prunus domestica, Prunus cerasifera, Punica granatum, Ribes rubrum, Sambucus nigra, Malus sylvestris, Cosmus ananassa, Tarassacus officinalis, Citrus synesis,* starchy *Oryza sativa, Ordeum vulgare* and *Zingiber officinalis*. With the help of the beekeeper Aristaeus would be possible also to produce mead, the wine made fermenting the fructose monosaccharide in the honey. Starch and

'Timeless fermentations' © Copyright Antonio Silvestro, 2019

pectin break down with amylase and pectinase enzyme is generally accomplish for increasing the fermentation ratio ethanol/sugars, otherwise, alcohol percentage.

The **grape vine** (*Vitis vinifera*) is native form Europe, Africa and Asia with currently up to 10000 varieties characterized by liana (32 m), palmate alternate leaves (5-20 cm), berry grape (up to 3 cm). Red wine is pigmented by the anthocyanins (cyanidin, delphinidin, malvidin, petunidin, peonidin-3-O-glucoside), while, the white colouring is lacking of this dying. Tannin are presents in their cuticles from which recycling could led to the generation of by-products suitable for vegetable leathers for clothes making. The skin brightening stilbenoids phytoalexin resveratrol (0.2 - 3 mg/mL) is expressed against physical damage, fungal pathogens *Botrytis cinerae*, or UltraViolet (UV) as exceptional anti-aging properties. The melatonin found in its grape is the chemical compound that induce lethargy.

Liquid chromatography:

- Stationary phase: potato starch, silica gel or sand
- Mobile phase: acetone, water, ethanol or their mix
- Column: syringe, pipette, bottle, tube with cotton on the bottom

Isolation and characterization of resveratrol from *Vitis vinifera* using reducing UltraSounds (20 < v_{US} < 100 kHz US 13 €, Sonic 45 € https://www.amazon.com/gp/cart/view.html?ref_=nav_cart) for t = 1 h preferentially using earplugs for avoiding their passage via the *Brain Blood Barrier* (BBB) corrupting the regular functioning of the auditory cortex reticular formation in the brainstem leaving the so-called 'jump scares' that would compromise the memory holding by the hypothalamus, separating the extract precipitated from the supernatant using SiO_2-chromatography, storing up to t = 2 months at T = 25 °C.

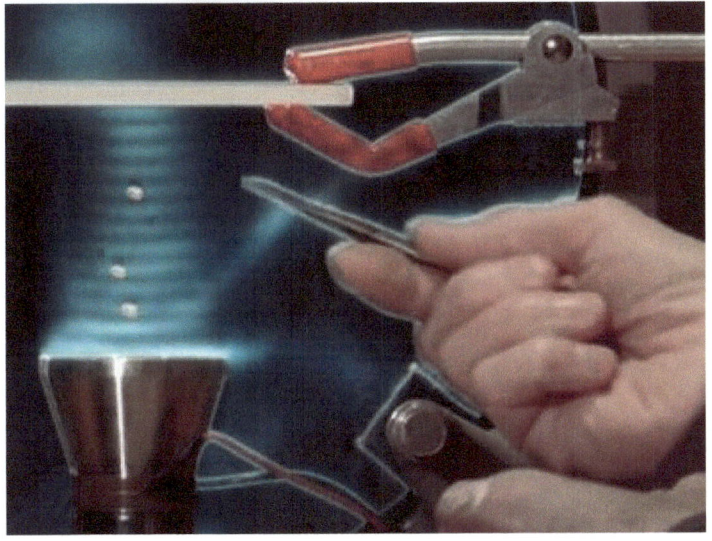

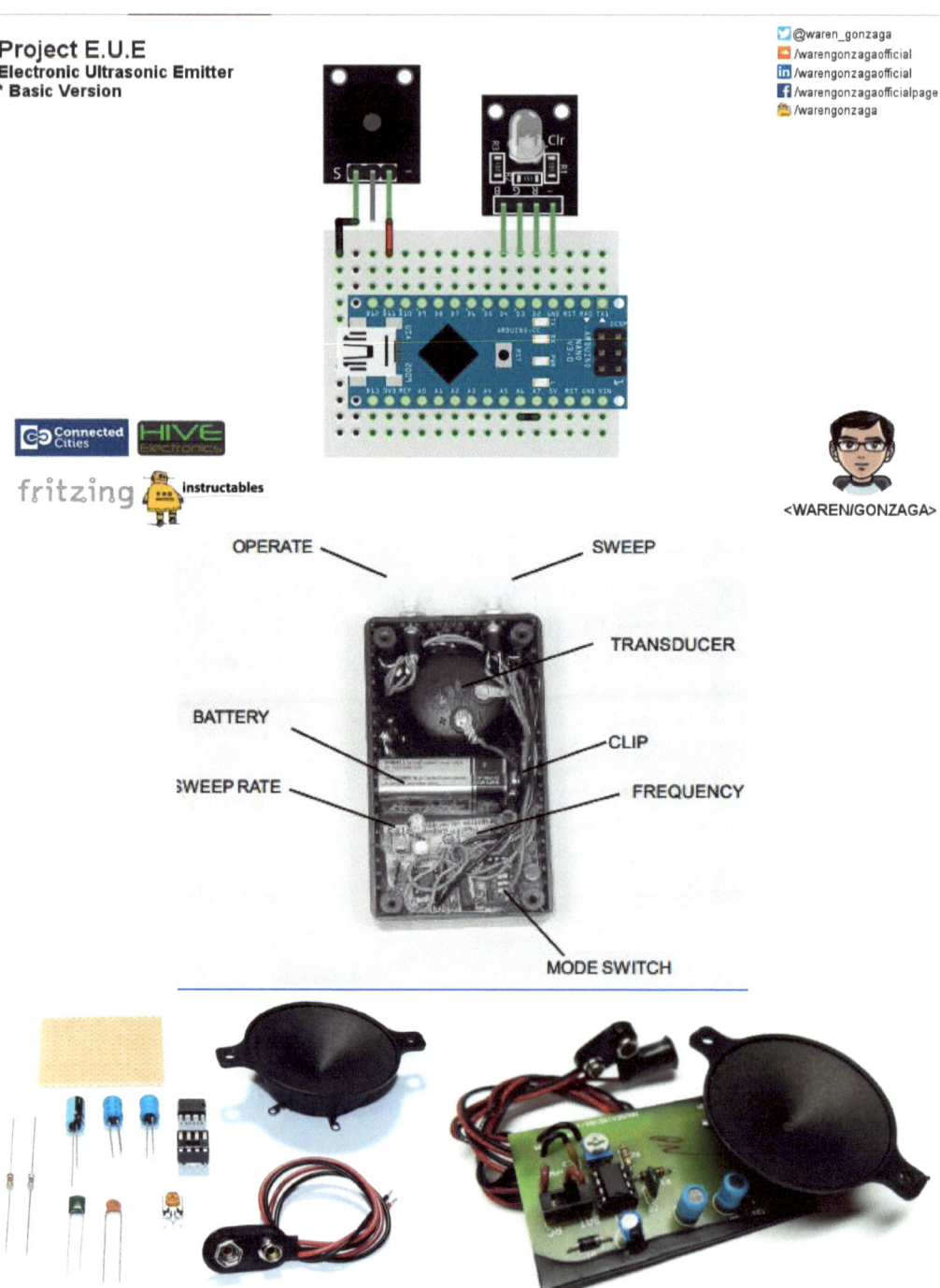

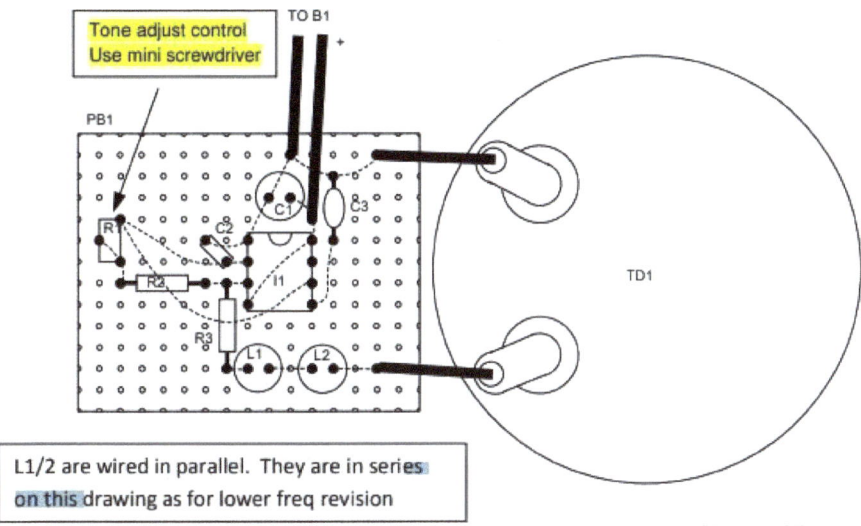

Figure 21 Ultrasonic power board: $\Delta V = 110$ V, P = 50 W (first row left), Pendulum transducer: P = 50 W, ν = 40 KHz (first row right). UltraSonic (US) levitation (second row). Arduino Nano and speaker piezo-electric buzzer (20 < ν < 65 kHz) edited adding a trim pot could led to his utilization with superior pests (third row). People more than targeted inferior species potential phytopathogens may be repelled by this frequency ν = 12 kHz, among the many bolts of Zeus (fourth row). US RCL generator based on NE555P timer, trim pot, switch, two resistors R = 200 Ω, two inductors, electrolytic C = 100 µF, ΔV = 50 V, ceramic 103 and polyester capacitors, LiPo (ΔV = 3.7 V, C = 2 Ah), alkaline (ΔV = 9 V), or 8 x AA battery (ΔV = 1.5 V), t = 6 h, $\varnothing_{speaker}$ = 5 cm, (P ≈ 100 W, d_{RC} = 900 m, SPL = 115 dB : d = 5 m, 80 €) (fifth row) – Image source: 36 € https://www.banggood.com/110V-50W-Ultrasonic-Generator-Power-Supply-Module-1pc-40K-Ultrasonic-Transducers-Vibrator-p-1400910.html?gmcCountry=IT&currency=EUR&createTmp=1&utm_source=googleshopping&utm_medium=cpc_bgs&utm_content=xibei&utm_campaign=xibei-pla-it-en-pc-all-0608&gclid=CjwKCAjw4_H6BRALEiwAvgfzqwkHxEnZq8pdKLUaViiktAi2bevqo2Zm-BXakqEuTUWNnhlPqY6HfRoCXMQQAvD_BwE&cur_warehouse=CN, https://www.youtube.com/watch?v=XpNbyfxxkWE&feature=youtu.be, https://www.instructables.com/id/Electronic-Ultrasonic-Emitter-Basic-Version/9, https://www.amazingl.com/products/sonic-nausea-device-with-radio-control.html, 'UltraSonic (US) generator' breadboard (9V battery) © Copyright Antonio Silvestro, 2020.

'Timeless fermentations' © Copyright Antonio Silvestro, 2019

Acetic fermentation - Vinegar

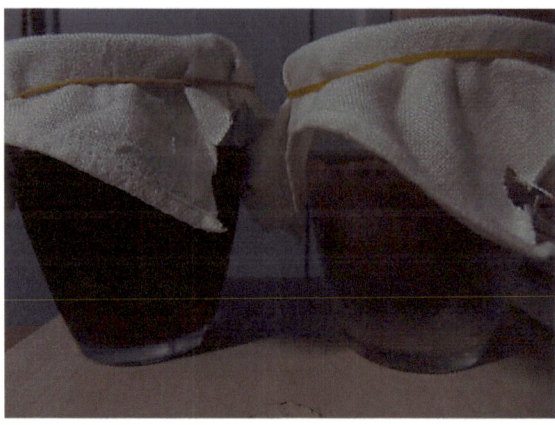

Figure 22 Cheap, old, low ethanol percentage wine even diluted into water (75 :25) can be easily converted into homemade vinegar, placing it into a flat tar 2/3 filled secured with cheesecloth and rubber band stored into dark and cool place, shaking once a week, for 2 weeks < t < 2 months – Image source: © Copyright Antonio Silvestro, 2022.

Vinegar used in the kitchen, as in the garden as annuals herbicide and pesticide, can be made by leaving one of the ten genre Gram-negative obligate aerobes Acetic Acid Bacteria (AAB) *Acetobacteraceae* (≈ 1 x 3 μm, e.g. *Acetobacter* (Bejerinck) *acetii* (Pasteur), 12.5 $ + US shipping htttps://www.m.carolina.com, https://www.microbiologist.com, ATCC® 23746 486 € https://www.igcstandards-atcc.org, *Mycoderma aceti* 9.9 €/0.2 L https://www.materiamadre.it), identifiable and isolable using 16S rRNA and PCR amplification form soil, fermenting organic vegetal substrates like grapes must, grains, coconut, pomegranate, apple cider and/or orange peels, at T ≈ 25 °C and pH = 5.5, oxidizing ethanol into acetaldehyde by a pyrroloquinoline quinine-dependent Alcohol DeHydrogenase (ADH) and then it into polar hydrophilic solvent acetic acid (concentrated between 4 < CH_3COOH < 8 %) by an ALdehyde DeHydrogenase (ALDH).

Sugars in various sources may be Capricorn fermented into ethanol, Virgo distilled (1h/L) for eliminating water in Vevor purifier [https://m.vevor.com/alcohol-distiller-c 10688/4l-home-countertop-stainless-steel-interior-water-distiller-purifier-machine-p 010862827194 - clean with citric juice or extract acid ($C_6H_8O_7$)], this poured into vector microworms (*Anguillole aceti*) batch culture for being transformed by their prey AAB into vinegar (recommended dose at 1 tbs /day).

The balsamic vinegar is generated cooking concentrated red grape must, that give it the typical reddish colour, sometime adulterated with the colouring agent E150d, and higher acidity (6 %) than the conventional one.

Figure 23 Roof-covered, open-air, manually agitated daily, EPS and glass bottle tank discharged of 1/3 balsamic mother acetic acid ($\rho = 1.1$ g/cm^3) and refilled of 1/3 ethanol ($\rho = 0.8$ g/cm^3) each month, harvesting the denser double fermented via a bottom faucet – Image source: © Copyright Antonio Silvestro, 2022.

Mercury in Virgo Distillation

Hot distillation: put filtered extract (or just water - deionization) in the ice box tray and let it freeze (t = 4 h, T = - 18 °C). Dripping distillation of the iced extract putted on the holed lid, in an upside-down glass jar. Distilled extract drip into a glass bowl, itself submerged in a metal bowl filled with ¼ volume of tap water as Baine Marie for t = 5 min, till the iced extract is melt (repeat with the felt liquid n-times).

Cold distillation (fractional freezing): freeze (T = - 18 °C) hydro soluble solvent in a bottle, then move it in the fridge (T = 4 °C) upside-down in a jar and let drip the water while the ethanoic acid (T = 16 °C) would remain solid ice, while, diluted solvent such as ethanol (T = - 114 °C) and acetone (T = - 95 °C) will be purified all in the freezer, leaving them drip leaving solid ice in the originary bottle, because the freezing point of the solvent is lower than water that become ice [e.g. ethanol ($\rho = 0.8$ kg/L) solvent isolation from freeze distillation of liquor beer (melting points: T_{H2O} = 0 °C and T_{C2H5OH} = - 114 °C)]. Let drip the ethanol n-times measure in each the alcoholic percentage increment with a hydrometer, but take care if using very small bottles (e.g., V_{PET} = 75 mL), because also water will melt at a ratio of 1 inch^3/h (in 3 h the ice in the PET would melt mixing again with ethanol in the saving glass – let it drop down for lesser time, for safety, 1/3 t = 1 h).

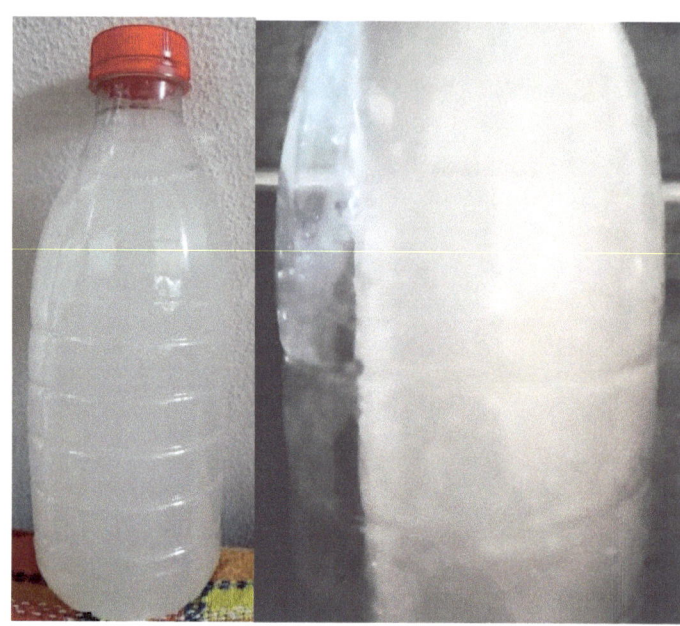

Figure 24 Kilju Finnish alcoholic beverage sugar wine, sparkling (1 < p < 3 atm), because of carbon dioxide (CO_2) accumulated while stored in the fridge at T = 4 °C (left). After the fermentation of Capricorn, Virgo would help you cold distil separating the icy water from the liquid ethanol of concentrations superior of the 15 %, placing the bottle horizontally in the freezer, obtaining V ≈ 150 mL for each litre of originary ethanol of lower content, suitable for disinfection also together with homemade soap (for more info *'Inanna cosmetics'* eBook 1.03 € https://www.amazon.com/dp/B08TV938PT) (right) – Image source: © Copyright Antonio Silvestro, 2021.

The alcolic percentage can be raised using an azeotrope like toluene or drying agents such as molecular sleeve (3A, 31 € https://scubastore.com), magnesium, calcium or copper sulphate (Mg, Ca, $CuSO_4$); this lasts obtainable reacting sulphuric acid (H_2SO_4 40 €/25 L https://www.lavanderiastore.it) with calcium carbonate ($CaCO_3$) egg shells.

An efficient way to use the Hercules serum from soft cheese curd separation is to use it for dissolving sugars fermenting into ethanol for obtaining alcoholic milky beverages.

<< Is still worth it? >> Would ask the maiden to the Sun, leaving it speechless to let Pluto wort be so valuable still, life after death rebirth with tears of Jupiter into which the moonlight would dissolve its sweetness.

Aries separate life from death, eat the first and drink last. Take care to not be intoxicated

'Timeless fermentations' © Copyright Antonio Silvestro, 2019

fermenting for example Solanaceae peels containing glycosilalkaloids, but uncompostable acids Rutaceae peels.

Timeless fermentation ♑

Rebirth in your liver and thyme, is against alcoholic beverages drinking, for which is needed to use it eating food like bread and meat.

Yeast Saccharomyces cerevisiae can be used as feed for Anguillola aceti, Panerellus redivivus, Gallus gallus, Oryctolagus cuniculus e Carassius auratus. It lasts 10 days, for which is advisable to feed it with sugars weekly into a fertmenter.

An efficient way to use the serum from soft cheese curd separation is to use it for dissolving sugars fermenting into ethanol for obtaining alcoholic milky beverages.

<< Is still worth it? >> Would ask the maiden to the Sun, leaving it speechless to let Pluto wort be so valuable still, life after death rebirth with tears of Jupiter into which the moonlight would dissolve its sweetness.

Aries separate life from death, eat the first and drink last. Take care to not be intoxicated fermenting for example Solanaceae peels containing glycosilalkaloids, but uncompostable acids Rutaceae peels.

36 €/ 3 L https://www.amazon.it/Quercia-Rubinetto-Dispenser-stoccaggio-Tequila/dp/B08JCV994F/ref=mp_s_a_1_22?crid=2J39LVR9FCGZF&keywords=botti+di+legno+3+L&pscroll=1&qid=1653248323&rnid=490259031&sprefix=botti+di+legno+3+l%2Caps%2C111&sr=8-22&wIndexMainSlot=26

3 L tank/ 8 € https://www.vipostore.it/it/taniche/39969-tanica-3lt-8032532520031.html

Virgo distillation Vallardi 50 L https://www.corriere.it/cronache/09_maggio_03/lorenzo_salvia_la_grappa_fai_da_te_diventa_leg ale_b0e057b0-37b3-11de-8d05-00144f02aabc.shtml#:~:text=Sanno%20come%20si%20fa%2C%20non,litri%20di%20grappa%20l'anno.

Pesticide done of ethanol produced from potato as following: liquefaction of potato starch beige slurry was done with salivar α-amylase at $T = 80$ °C for $t = 45$ min followed by saccharification process which was done with glucoamylase at $T = 65$ °C for $t = 2$ h. Fermentation of hydrolysate with Saccharomyces cerevisiae at $T = 35$ °C for $t = 2$ days resulted in the production of $ro = 33$ g/L ethanol.

www.ingramcontent.com/pod-product-compliance
Lightning Source LLC
Chambersburg PA
CBHW040245220526
45473CB00001B/375